Food Roofs of Rio de Janeiro

Rob Roggema
Editor

Food Roofs of Rio de Janeiro

The Pavao-Pavaozinho and Cantagalo Case Study

Springer

Editor
Rob Roggema
Faculty of Design, Architecture and Building
University of Technology Sydney
Ultimo
Australia

ISBN 978-3-319-56737-2 ISBN 978-3-319-56739-6 (eBook)
DOI 10.1007/978-3-319-56739-6

Library of Congress Control Number: 2017938131

Printed on acid-free paper

This Springer imprint is published by Springer Nature
The registered company is Springer International Publishing AG
The registered company address is: Gewerbestrasse 11, 6330 Cham, Switzerland

Foreword

In 1897, Brazilian author Euclides da Cunha described a 'mount favela' in his book about Brazil's military campaign to suppress civil upheavals in Canudos. Soldiers camped on a hill where the thorny favela plant grows in the north-east region and made temporary housing out of shacks. When these soldiers returned to Rio de Janeiro, they settled on the hillsides of the city, waiting to be granted the land the government promised them. Their living conditions reminded them of the 'favela' hills. The term eventually stuck. Today three quarters of all Brazilians live in cities. In Rio de Janeiro, Brazil's second largest city, almost one quarter of the population, more than 1.5 million people, lives in these so-called favelas.

Within half a century, Brazil has changed from a rural society into a predominantly urban one. The increase of urbanisation has put an enormous pressure on housing and public services, such as water and sanitation, health care and education. Although Brazil has had some success in reducing poverty and inequality over the last two decades, the problem of social inclusion in society is still very much present. This is probably nowhere more visible than in the favelas situated on Rio's hilltops, so close to the postcard-like beaches and luxurious condominiums.

In 2013, on one of his travels, Dutch professor Rob Roggema came to Rio and got acquainted with living conditions in the favelas Cantagalo and Pavão-Pavãozinho. When he visited me in the consulate and told me about the FoodRoof, I was surprised by the simplicity of the solution yet the challenges ahead.

First and foremost, the FoodRoof solves one of the most basic problems of poor people: the access to healthy food. This time, the problem is not approached from a patronising perspective of sending food support to people in need, but—in line with the old adage of development cooperation 'teach a man to fish and you feed him for a lifetime'—it aims to give the people the tools in hand to grow their own healthy food.

The FoodRoof project is a good example of connecting theory and practice. One of the problems of our time is that we invent highly sophisticated solutions for all kinds of problems, but these are often not put to practice. Especially when the needs of the poor are at stake, it is essential to develop projects in which these theoretical solutions are realised in practice, with the involvement and ownership of the beneficiaries themselves. The FoodRoof represents such a participative approach. It connects the global problems of food security and food safety with a simple, effective local solution.

In the Netherlands, one of the most densely populated countries in the world, available space is limited. Competition for land in cities is high; various social and economic functions have to be reconciled. The same can be said about the conditions in the metropolitan area of Rio de Janeiro. Space for horticulture and agriculture is limited. The global movement to develop urban agriculture is certainly 'hot' in the Netherlands, and it is encouraging to see that our gained knowledge and experience is shared and tested elsewhere, in this case in Brazil.

At the same time, the realisation of the first FoodRoof also illuminates how local circumstances and limitations influence the possibilities to realise these kinds of projects. The project manual, translated into Portuguese, has proven to be very helpful in getting the purpose and practical guidance explained to people in the area. However, the follow-up, after the first FoodRoof was realised, could have been more extensive. It is regrettable that up to now—due to financial and political reasons connected to the current crisis in Brazil—not more of these FoodRoofs were implemented in Cantagalo and/or in other favelas across Rio de Janeiro and beyond. Once a second and third FoodRoof would have been realised, the prototype could be improved as the teething troubles of such a project were addressed.

The FoodRoof shows a big potential in solving local people's problems, and at the same time giving them the responsibility for making sure the FoodRoof system is operating well and the crops can be harvested as they should. A successful FoodRoof gives families in the favela the opportunity to provide their children, and themselves, with healthy food.

This book gives an insight in the problems with food of our times and the way it can be practically approached. Dutch inventiveness has been coupled with the Brazilian art of improvisation. It has brought together international and Brazilian experts and expertise, local knowledge and insights, and the practical skills of Dutch students and Brazilian professionals. Working together, learning together. Out of this combination, this showcase of a FoodRoof emerged. By far a single roof is not enough to solve all the problems in the favela, but it is a hopeful sign of progress, which deserves imitation and replication all around Brazil and beyond.

May this book inspire you and mobilise partners to not only theoretically debate the FoodRoof idea, but also to apply this thinking to many other contexts, inside and outside the favela.

Arjen Uijterlinde
Consul-General of the Kingdom of the Netherlands
Rio de Janeiro, Brazil

Contents

Introduction

On travel through South America we once visited Rio de Janeiro. We met up with Amaury Pugliese who showed us around the beautiful highlights of the city, but brought us also high up in one of the favelas: Cantagalo. On a dark evening in August 2013, we visited a jiu-jitsu school for kids in the favela. Around 40 kids, around twelve years of age were training, every evening, for approximately 2 h. No wonder the best kids are Brazilian champions in their age groups and dreaming of the Olympics. This project is one of many to socialise the favelas: this one in particular to keep young boys on the right track, teaching them the importance of respect, order and discipline. On our walk down the hill, I asked Amaury what we could do to help these kids from a distance. He told me it would be great if we could send a bag of beans or a bag of rice every month. They lack food in its most basic form. This ignited the FoodRoof. If we could teach them how to grow their own food, the necessity to send bags up Cantagalo hill would be gone.

Imagine, to have lived in Cantagalo since you were born. The daily routine comprehends playing games and searching for your parents, if they are still alive. Hoping to stay away from drugs trafficking, but also attracted to the excitement and its golden promise. By the end of the day you are feeling hungry so you try to find a snack, some chips and lemonade to feed you. Or you try your first booze and get dependent on 'others', whoever they may be, as they have some kind of power over you. These people provide a safety net of sorts. But you're still stuck with old food habits, such as alcohol, pre-wrapped cakes and candy. I met Rafael Lezinho. He is a tough guy, nearly two metres tall and one of the driving forces of the Team VB Jiu-Jitsu School in Cantagalo. I visited him in his home, and he showed me around the favela. I asked him what he eats and besides some pre-wrapped sandwiches, cakes and candies, it wasn't much. I wondered how he could stay alive. At 21 years of age he already weighed over a hundred kilos and trained each day three times to get rid of all the Kilo-Joules he gained during the day. It was clear to me that he could use some vitamins and fresh food. There was only one solution: start growing his own food in his own neighbourhood, on his own roof.

A FoodRoof?

The FoodRoof supports residents of the favelas in Brazil to grow their own food. A FoodRoof is a closed aquaponic system, producing fish, vegetables, fruit and herbs, is lightweight, is easy to construct and fits on a (small) roof. With more than 50% of the world population living in the city, and roughly another 50% living with limited access to a healthy diet, the question is no longer if we can produce enough to feed the world, but how and where we must grow food to feed everyone. The answer could well be that we shouldn't become more efficient, large scale and more productive, but that we need to grow food in places where people suffer from eating unhealthy food. Instead of educate people to start eating more healthy food, we need to teach them to grow it by themselves. In order to do this, we need to enter deep into these communities and create productive city-landscapes. This is what happened in Cantagalo.

Cantagalo

When standing on Ipanema or Copacabana beach nothing reminds you of the favelas Cantagalo (Fig. 1), which are hidden behind tall luxury buildings along the coast. But in reality, the heart of the favela is only a 10-min stroll away. Uphill the seemingly randomly built houses follow the pattern of the slope. The streets have followed the topography in well-organised yet insurgent patterns. The northern side, also the upper ridge in the area is forested and new housing initiatives nibbling their way into this space bit by bit. On the other sides of the favela, the typical harsh

Fig. 1 Aerial view of Cantagalo and Pavão-Pavãozinho. *Source* Google Earth

and sudden transition to luxury apartment buildings and pet-loving citizens can be experienced. One moment you walk in the middle of a buzzing and mindboggling emergent environment in between all kinds of houses and shops, the next you find yourself in the middle of large and secured houses of the stiff upper lip.

The houses in the favela are built with whatever material was available or could be obtained for free. The strength of the buildings is generally a concern, especially because materials are used that happen to be around at the time of building. A construction drawing was often not in sight. However, there is a very good reason to trust the construction safety and the strength of the buildings. The main incentive for constructing strong buildings is the fact every owner is very keen to rent out an additional floor, simply because of earning money. Therefore, the constructive strength of each house must be suitable to build additional floors. This created a rich variety of buildings housing, some higher than others, but all of similar sizes and with flat roofs, or the opportunity to create one. The latter is important, as spaces are small and densities high (Fig. 2). This means the amount of open, available and green spaces is limited and spaces for growing food likewise. Roofs seem to offer opportunities here and this has been identified, though slightly hidden, by local artists as well.

The main problem of these areas is not the threat of violence, or fights with the police, it isn't even joblessness. It is a lack of sufficient and quality of food. Many inhabitants do not have direct access to healthy food, because it is only available in the bigger supermarkets, where prices are (too) high. Local shops only sell candies, other stored goods, and no fresh vegetables or fruits. It is not possible, convenient

Fig. 2 Diversity of houses with flat roofs. *Photo* Rob Roggema

or easy to bring these products from outside into the favela. It may be too expensive, it does not stay fresh in the heat, and it may be not safe for the traditional food providers from outside the favela. This requires an alternative approach, in which inhabitants have a say about their own area, turning it into a productive space and trade the products locally.

This book describes the development of the 'FoodRoofRio', a way of providing food to the slums of Rio. In Chap. 1, the role of food in spatial planning and design is described. Chapter 2 continues with an analysis on the spatial requirements urban food production has and Chap. 3 digs into the networks and chains that are necessary to distribute food from local growers and producers to local consumers.

In Chap. 4, the policy context for upgrading the favelas in Rio de Janeiro is described from a governmental perspective. Chapter 5 deals with ways to engage local residents in the design process. In Chaps. 6–9, the design and realisation is presented in Chap. 6, the overall design strategy in Chap. 7, the process of how the FoodRoof could be realised, the detailed design in Chap. 8 and how the first FoodRoof has been built in a period of one week (Chap. 9). In Chap. 10, the owner of the house on which the FoodRoof has been realised is interviewed and Chap. 11 ends with a conclusion and outlook.

Part I

Chapter 1
Food in Spatial Planning and Design

Rob Roggema

1.1 Introduction

Food safety and food security are major topics in the current global debate about food production and provision. The availability of food is a major issue and many people focus on productivity and efficiency in agricultural practice to deal with these issues. However, to know if your food is safe, healthy and available, the distribution of food over the world is more important than concerns over the amount of food produces. Worldwide, the food production can be enough to feed 7, 8 or 9 billion people, depending on the policy decisions taken in the next few years. If we increase yields and expand crop area we could produce more, if we close the yield gaps, if we reduce or improve meat production and if we reduce food waste we will be able to keep on feeding the world population (Hofstrand 2011).

As it comes to distribution of food, we might look at this from different scales. At the global level, this is a logistical and financial problem. How many and how much does it take us to send containers with crops around the globe? This, more a traditional, economic and expansion-driven perspective, is placed against a more local oriented way of distributing food. Regional markets providing food for their own population do not require a huge transportation system, and makes it easier and more flexible to match demand and supply. Other environmental benefits are obvious, such as a better control over the quality and ways of producing food, lower food miles and a higher involvement of citizens in the food production process. This alternative is still far from real and many systems need to change before the food system can transform. But the traction is there and people are becoming more aware the mass production of food might be not the best option for our future food system.

R. Roggema (✉)
Faculty of Design, Architecture and Building,
University of Technology Sydney, Ultimo, NSW, Australia
e-mail: rob@cittaideale.eu

R. Roggema (ed.), *Food Roofs of Rio de Janeiro*,
DOI 10.1007/978-3-319-56739-6_1

Still, even if the food production is redistributed and more localised it is not a guarantee on having access to basic food for the people who are most in need of it. The distribution of food over different social groups within communities is all but solved. The question here is how could food be provided for the poorest people in the slum areas around the world? To achieve this, access is the key variable and the distance to food both in (kilo)metres as in price is important. When people are close to where the food is produced and the price is low enough, they will have easier access. Here food production directly interferes with spatial planning and design. Especially in urbanised areas, more specifically in slums, space is limited and there is a high competition between the different land-uses that have to be accommodated. To find a place where food can be grown requires smart design.

In this chapter the way food is part of spatial planning and design is explored, a framework for the design of urban agriculture is presented and guidelines are given how to improve the role of food in the urban context.

1.2 Food in Current Planning Theory

The ambition to create productive cities is generally seen as a beneficial, sustainable idea, contributing to food security, health and social cohesion. It seems logical to grow food close to where it is consumed. With over 50% of the global population living in cities, the city should be the prime location to produce food. At the same time, an increasing number of people live in squatters, some temporary, but most of them permanently. In these areas it is often very difficult to supply enough and healthy food. Moreover, distance is a factor of importance. If food miles could be decreased the environment and freshness of products benefit, and it also increases the knowledge about and security of food.

Current planning and design theories in which agricultural activities play a dominant role are mostly oriented on the countryside. In the Dutch tradition for instance, designs are based on the so-called 'casco-concept' (Sijmons 1992) and/or 'the strategy of the two networks' (Tjallingii 1993). These two theories separate high dynamic land-use from the low dynamic ones.

On the other hand, in urban design theory, food production is hardly an issue because the spatial layout predominantly focuses on the diversity of land-use, allocation of residential areas, the economic and transport systems, and multi-usable public spaces. It is clear that the current highly productive forms of agricultural practice, which are mono-functional, make use of large-scale precision methods, use intensive breeding methods, consist of large scale greenhouses and open field cultivation, face difficulties to fit their spatial, environmental and societal characteristics in the spatial patterns of current urban environments, which are often the opposite: high densities, small public spaces, multifunctional and intense types of use. The competition for land in cities is (too) high to easily include agriculture.

Additionally, in current Urban Agriculture practice three problems coincide. First, most of the realised projects emerged out of 'local good will'. An active person in the neighbourhood thought it could be a good idea to grow some food in the immediate surroundings of his or her house. Many projects were initiated this way and the lucky ones survived. In this case, lucky stands for the fact the initial initiator of this project hasn't moved to elsewhere or got disinterested in growing food.

The second problem many projects face is they are in one way or another supported by external funds, which are, in the majority of the cases, temporary. These projects often receive subsidy from the government, which could make their existence on the long term problematic.

The third problem is the lack of design. Many urban agriculture projects are hardly designed, are located alongside infrastructure (rail-tracks are prime) and out of sight of the ordinary user of the urban environment. Often there is a good reason for not showcasing urban agriculture: it often lacks structure, is badly connected with other places and could look ill maintained. Most of these projects are located in peripheral places, occupy land that is temporarily not used, are hardly designed and seem to be developed without much thought about their appearance. When urban agriculture projects are realised without thoughtful planning and design it does not increase the acceptability nor popularity. There might be a crucial and underestimated reason for all this. Despite many people think urban agriculture has been around for centuries, it is actually a recent phenomenon. The idea of monks gardening in their cloister gardens, food production within walled cities in the Middle Ages or growing fruit and vegetables in renaissance gardens is all very romantic, but it didn't take place in cities. Even the Volksgarten, Allotment gardens or Moestuinen were mainly created to offer busy and stressed city people, living in small dense and poor parts of the city, some free green space to spend the weekend. These gardens were primarily meant for leisure rather than to produce food, originally located at the city fringes, which only became included in the urban environment over time.

1.3 For Long, Urban Agriculture Didn't Exist

In history Urban Agriculture was no part of urban design or city development. When the growth of food in cities is to be taken seriously, it should imply to arrange a prominent position in the planning and design discourse. However, when similar design concepts and principles are used to design productive cities as they have been applied for non-productive ones, they won't enhance the productivity of the city.

City concepts that have been developed over the years show clearly where the agricultural production was projected. In Von Thünen's model (Von Thünen 1826) for instance (Fig. 1.1) it was clear that all types of agricultural activities were found in concentric zones around, and all outside, an urban core. The freshness of products determined their respective distances to the centre.

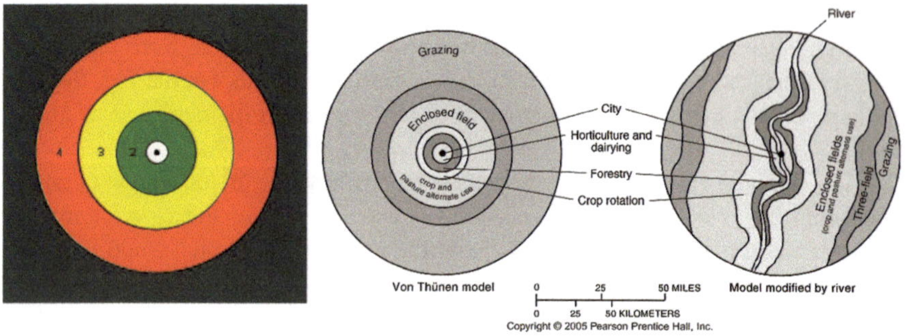

Fig. 1.1 Von Thünen's Model with from the center outwards: CBD, horticulture (dairy, fruit, vegetables), forest, crops (grain), grazing livestock, Wilderness (Von Thünen 1826)

Howard (1902) locates large farms in specific sections between the central city and the satellites in the countryside, and positions allotments around some of these satellites (Fig. 1.2).

La Ville Radieuse (Corbusier 1935) proposes a zoning of the urban environment in a way Von Thünen would have loved and locates agriculture in the outskirts of the city (Fig. 1.3).

In Broadacre city (Wright 1945), a large amount of green spaces are introduced in the urban environment. The agricultural production is foreseen within the boundaries of Broadacre city, however limited to a small section of small farm in a leisure and recreation zone along the river and close to the market (Fig. 1.4).

These examples illustrate the way food supply was considered: as a source that needed to supply the city, but not as a part of a 'productive' city itself. As the analysis of Steele shows (Steele 2008), this hasn't changed over the years. Even when railway lines, industrial production techniques and urban development occurred, food production aligned with the developing urban environment (Fig. 1.5).

However, this poses a question: where in the city can Urban Agriculture be located? Food production has obviously not been a major factor in the design of cities, no matter what the designers would want you to believe. It has never existed as such. It must be framed as a new spatial planning, landscape architecture or urban design question. Are the concepts we are using to currently design our cities suitable for the design of Urban Agriculture? Or does Urban Agriculture face similar problems as to design a sustainable city? And if so, unsuccessful attempts to design a sustainable city are most probably not a good example for designing Urban Agriculture.

Still, there are designs that may inspire the design for Urban Agriculture. For instance, the green neighbourhood in Brøndby, Denmark (Fig. 1.6) showcases the spatial power of creating green spaces in between housing estates and is still extremely coherent. Imagine, these green spaces would be used to produce foods of all kinds. This spatial concept can carry this shift in land use easily and it could supply most of the required vegetables and fruits to the community. It might even

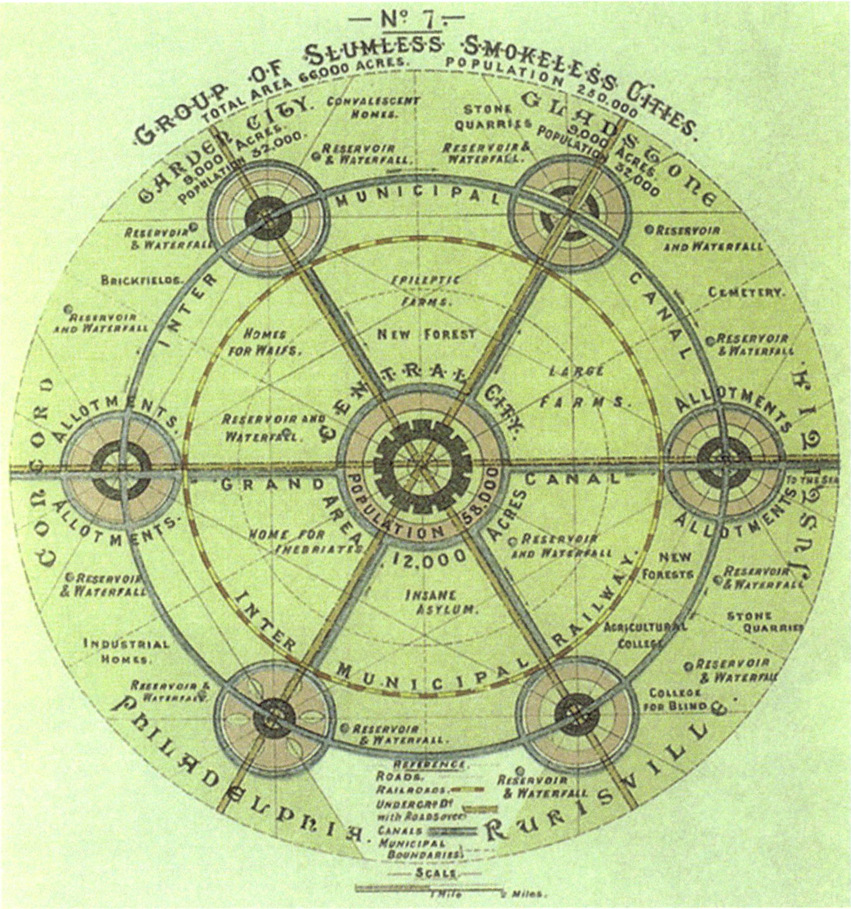

Fig. 1.2 Ebenezer Howards' scheme for his Garden City of Tomorrow (Howard 1902)

be a socially binding pathway, as residents could easily work and relax in these gardens and socialize. At the same time, the example illustrates the necessity of developing new spatial concepts. Concepts that will transform the city into attractive and productive places, where people, whether they want to be part of the food supply chain or not, are appreciating these productive environments.

Some of these new concepts have already been developed (but never realised). The Park Supermarket design (Fig. 1.7) is innovative as it sets out a clear new strategy in food supply, at a scale that holds the middle between the global food system and the local idealistic projects. It proposes to transform the landscape between Rotterdam and The Hague into a food producing supermarket, consisting of departments that produce vegetable, fish, meat, grain, spices and everything else you would normally buy in the supermarket, completed with a cash desk. It offers products from all over the globe, just like your own supermarket would. Only, it

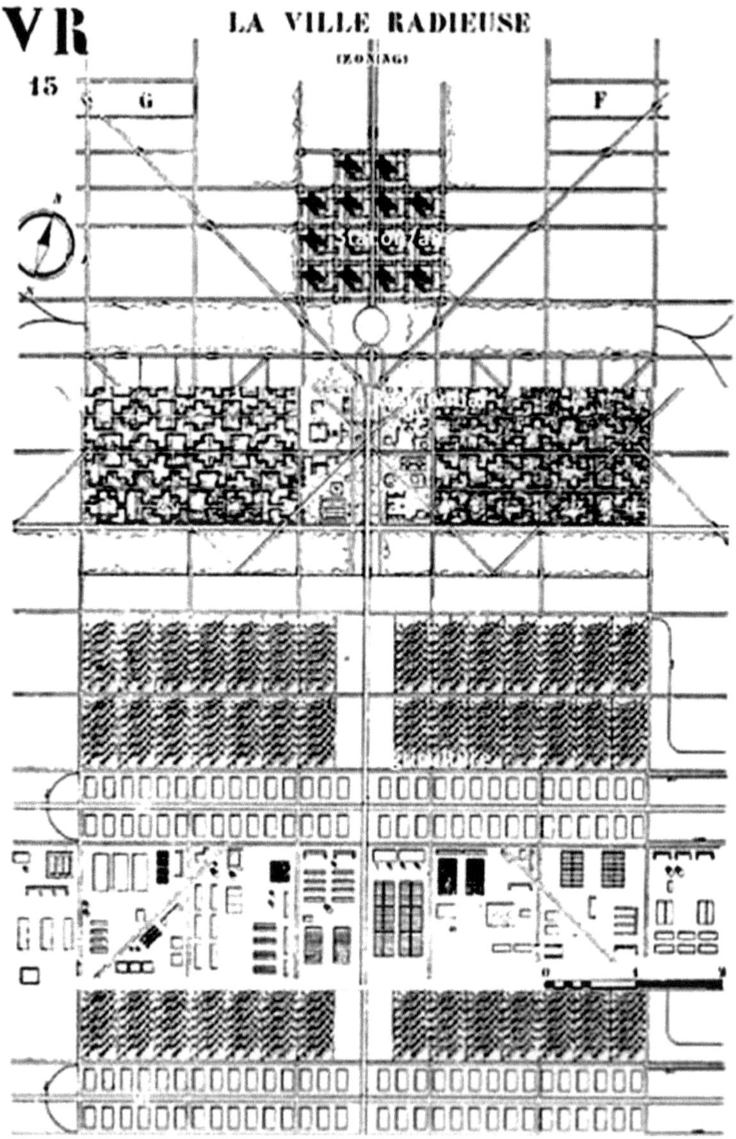

Fig. 1.3 La Ville Radieuse: from *top* to *bottom* the zones of University/Government, CBD, Station/Airport, Residential, Industry and Agriculture (Corbusier 1935)

provides the, sometimes artificial, climatic environments, created sustainably, in which these crops would normally grow. The design creates a landscape that could be interesting, not only for food production, but also for leisure, ecology and water conservation purposes.

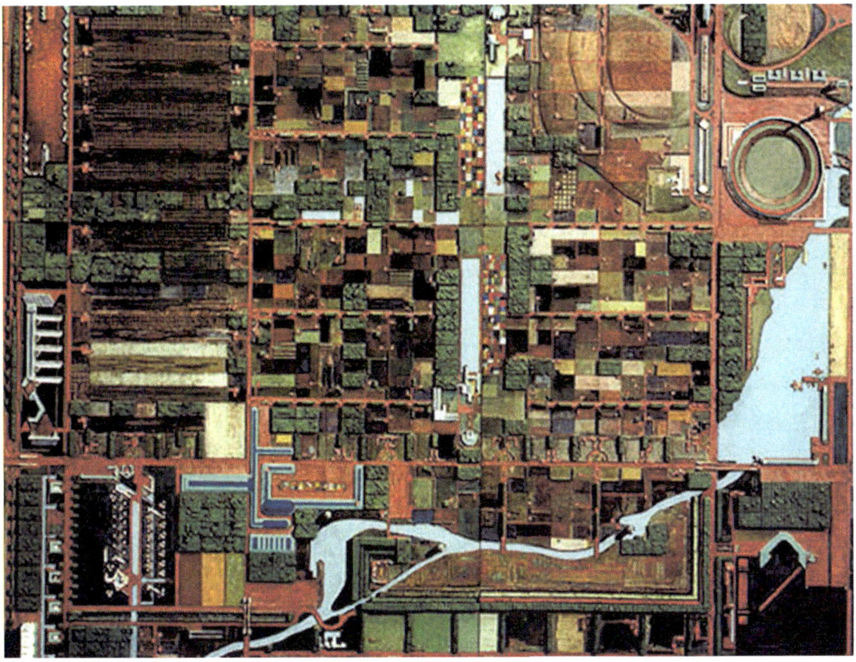

Fig. 1.4 Broadacre City (Wright 1945)

Fig. 1.5 Historic development relation between food production and the city (scheme by Steve Swiggers, after Steele 2008)

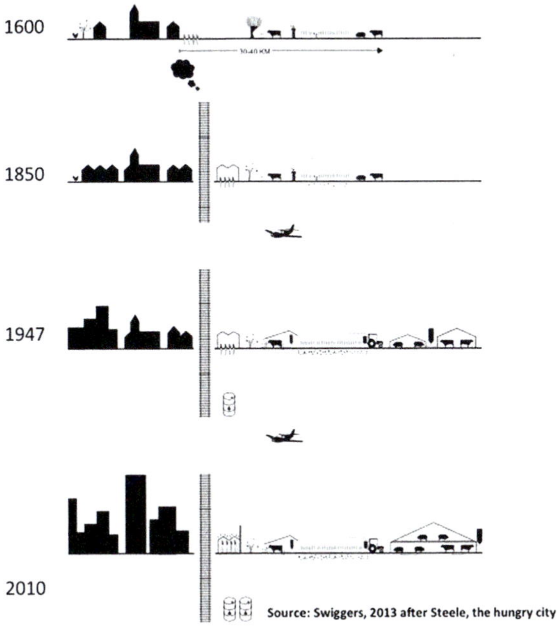

Fig. 1.6 Community of Brøndby, Denmark

Another innovative, but also controversial, concept is the design for Pig City (http://www.mvrdv.nl/projects/181_pig_city/#). This proposal is an attempt to solve three problems at once. In the context of the spatial density of the Netherlands relative big pig farms put spatial pressures on the landscape. There are simply (too) many pigs and the environmental impact of pig farms is huge. Pig Cities (Fig. 1.8) are proposed for the fringes of harbour areas, such as the Maasvlakte in Rotterdam, and consist of high-rise buildings, housing pigs in very low densities. Each couple of pigs share their own 'apartment', complete with balcony and outdoor space. They eat locally produced food and garbage from surrounding restaurants, and cuddle in their specially designed muddy living rooms. Their manure is collected and transported to a local energy plant recycling their waste as heat and electricity back into the building. This concept performs better for the environment than conventional pig farming, the pigs even feel better, but it has never been accepted as a realistic model by farmers and politicians as it soon became bon ton to reject these 'huge super farms', as they were seen as destroying the beauty of the landscape.

Interesting concepts have been developed, but what has been missing is a strategy that brings coherence, is acceptable for local leaders and residents, and connects interesting places with efficient food supply and environmental benefits. This requires Urban Agriculture to understand how the different foci, aspects and purposes relate to each other, allowing for the development of designs at every level, with an objective to transform cities into productive landscapes.

Fig. 1.7 Park Supermarket (Van Bergen en Kolpa 2010)

1.4 Towards a Framework to Design for Urban Agriculture

Given these problems, improving the coherence and multi-scaling is necessary to grow food in urban environments. Many aspects interfere, interconnect and interact. The framework (Roggema 2014a, b) presented in this chapter (Fig. 1.9) offers a chance to position all these aspects, to describe the way they relate, and to be of use in the design at the regional, city, neighbourhood and building level. In this framework two worlds collide: the functionality of the Urban Agriculture system (left hand side of the framework) and the way this system is designed

Fig. 1.8 Pig City by MVRDV (http://www.mvrdv.nl/projects/181_pig_city/#)

Fig. 1.8 (continued)

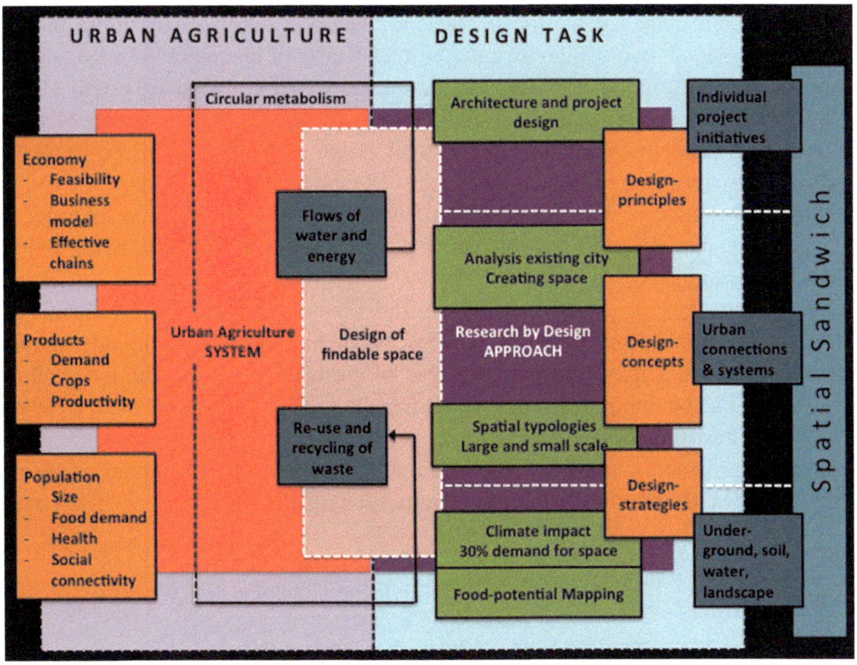

Fig. 1.9 Framework for the design of Urban Agriculture (Roggema 2014a, b)

(right-hand side). At its core these worlds come together, where a platform is proposed for integrated, future-proof designs of findable spaces.[1]

Factors such as economics, population and products determine to a large extent the urban agriculture system from a functional, socio-economic perspective.

First of all, the size of the population and the type of food demanded gives insights in the amount of food that has to be supplied. Moreover, issues as the objective to consume more healthy food or social aspects of growing food, such as working conditions and collaborative forms of shared productivity determine the sort of food demanded by a certain group or population. The main question is: "how much food must be produced to create a healthy and social attractive supply?"

Secondly, the crops themselves are a possible limiting factor in providing the desired products. Is there enough fertile soil, where the demanded range of crops can be grown? Is the productivity of the land high enough to grow a reasonable amount of crops?

The third aspect of the urban agriculture system is economic. Is the production of enough, differentiated and healthy food economically feasible? Can we develop business models that are profitable and can we design sustainable chains for products to reach the consumers in a safe, affordable and fresh way? These three basic questions for the urban agriculture system interfere, but may also profit from the circular metabolism model. In this model, the objective is to close the cycles of water flows, as well as those of energy, nutrients and materials and at the same time use as little resources as possible en use them efficiently. Moreover, the rest-flows are reused and recycled within the system. This offers urban agriculture projects several benefits. It could align the rest-products/waste of one place with the needs of another area. Secondly, it could create a coherent and connective system of food production network across the city and finally it could add up the volumes of water, energy and nutrients of a number of small urban agriculture sites, co-using and exchanging resources. The environmental benefits speak for themselves.

The other half of this framework consists of the design tasks, focusing on creating the spatial conditions for realising a spatial quality when growing food in urban environments.

First, the distinction between design and planning is interesting. Planning is defined as 'A basic management function involving formulation of one or more detailed plans to achieve optimum balance of needs or demands with the available resources. The planning process (1) identifies the goals or objectives to be achieved, (2) formulates strategies to achieve them, (3) arranges or creates the means required, and (4) implements, directs, and monitors all steps in their proper sequence' (http://www.oxforddictionaries.com/definition/english/planning) or as 'The control

[1]'Findable spaces' is a concept of thought, with the objective to find or allocate spaces in (dense) urbanised areas where many spatial constraints occur and a high spatial and functional pressure is experienced, for the production of healthy and fresh food.

of development by a local authority, through regulation and licensing for land use changes and building' (http://www.oxforddictionaries.com/definition/english/planning). Design is defined as 'A plan or drawing produced to show the look and function or workings of a building, garment, or other object before it is made' (http://www.oxforddictionaries.com/definition/english/design). Hence design and planning are related but different. To design is 'to plan and establish form' (De Jong and Van der Voort 2005). When design is used in a research context it can be defined in several different ways:

- Research for design is all the research required for making a design (often quantitative analysis of factors, relevant for the design and the location/context).
- Research through design aims to generate knew knowledge through varying the design solutions within a well-known context (http://boundariescrossing.wordpress.com/2012/07/17/research-through-design/).
- Research by design generates knowledge and understanding by researching the effects of changing the design solutions and/or the context at the same time (De Jong and Van der Voort 2005).
- Design research describes and analyses existing designs (e.g. comparative study) (De Jong and Van der Voort 2005).

Research by design is interesting when innovative designs are demanded and when the mutual impact on context and design are to be investigated. As discussed before, the field of research, urban agriculture is relatively new. Both the subject of the design, urban agriculture, as its impacts on the urban environment (the context) are objects of research. A research by design approach is very suitable as it includes the possibility to generate modifications and reciprocal changes of both design and environment.

1.5 The McAg

The scale at which urban agriculture takes place is generally located outside the city at higher spatial scales. Recent developments illuminate the development of urban agriculture projects at the lower spatial scales (see for instance: Viljoen 2005; Miazzo and Minkjan 2013; Philips 2013; Gorgolewski et al. 2011). Connections between the two levels are not very common. The higher scale is mainly represented in conceptual planning products, such as Master planning, regional planning or structure planning, while the lower scales are predominantly found in concrete project initiatives. In order to overcome the differences between scales and to realise beneficial exchanges between scales a spatial sandwich, the McAg (Fig. 1.10), has been introduced (Roggema 2014a, b). The design task focuses on connecting the scales that are part of this McAg: the individual, the urban and the regional scales.

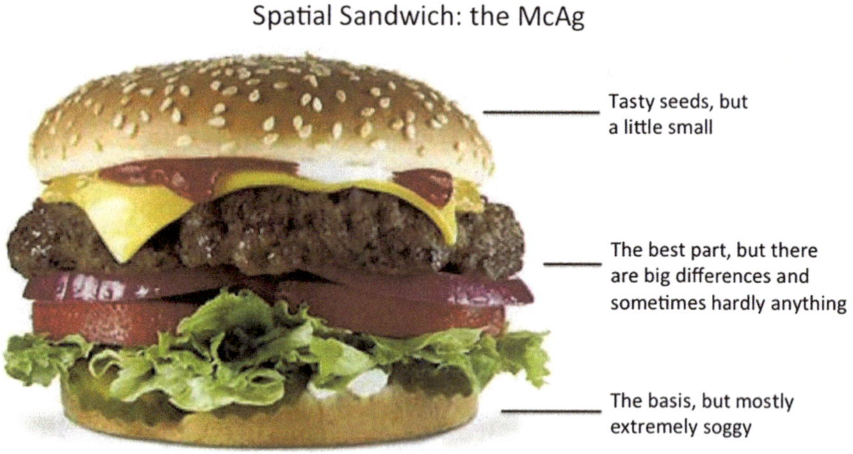

Fig. 1.10 The 'McAg' including three levels of the Urban Agriculture system, which are all necessary to create a productive city (Roggema 2014a, b)

The McAg is a metaphor for an urban system of design in levels including the connections between these levels. Urban agriculture requires design at three levels, including the design of the connections. At the upper level (The Top) of the McAg the seeds are found. The bun itself is dry and relatively big, but the seeds provide a nice taste. The only problem is, they are quite small. Similarly, this is often the case with urban agricultural projects. As little projects in the urban environment they're quite nice and the tasty products out of it are great, but these projects are small, especially in comparison with the large scale of the entire city. Therefore, these projects need to be designed in conjunction with the design of the middle level (The In-Between) where the best parts of the McAg are found. In between the bun, depending your choice of burger, the taste is created. Cities can only become productive (= tasty) when water, nutrients, the right soil and energy can reach places where food is grown. This requires careful design especially under influence of climate impacts. The specific combination of ingredients at the middle level guarantees a suitable and tasteful urban productive environment. The bottom (The Basis) of the McAg is the well-known soggy bit. It is not very tasteful, but necessary to provide the minimal strength of the whole. In the urban landscape this level is represented as the soil, the topography, and groundwater that together create the potentials for food production. Both natural as artificial landscapes can create the required basis for the middle and upper parts of the McAg. Again, at the underlying level, the landscape design could make or break the availability of productive conditions. The three levels of a spatial sandwich (individual projects, urban connections and the underlying landscape) are not necessarily seen as spatial scales, but coherent groups of spatial elements that together form an entity. However, when each of these levels is designed as a single issue the cross benefits are minimal and each of them can be easily neglected.

1.5.1 The Basis: Design Strategy

The underlying landscape is always there. It changes very slowly, if at all. This means we need to be very careful to use or extract resources from this system. To recycle the flows of water, energy and nutrients back into this underlying landscape must be a condition 'sine qua non'. Strategic planning is important here, in order to make sure that resources are available over longer periods of time. The scale of this underlying landscape is endless. It plays a role in defining the conditions at the very local scale, but is also important in connecting the local area to the entire landscape, and the urban space with the rural surroundings of cities. In order to keep the underlying landscape functioning, a long-term spatial strategy is required which emphasises the sustainability of the system and provides the spatial certainty for the landscape to maintain its functions. A smart 'metabolistic' system of supply, re-use and storage of resources is required to safeguard the sustainability of the underlying landscape. At this level a design strategy is most useful.

The 'soggy' base of the sandwich may not be the most tasteful, without a basis it would be hard to grab and lacks any strength.

At this scale the potentials of the landscape are important to map. Mapping a diverse set of food-potentials (FPM), similar to the way energy potentials are mapped (EPM, Van den Dobbelsteen et al. 2007) will give insights in the suitable or unsuitable places in urban regions to grow certain crops, or the type of food system that best can be applied. Aspects such as the soil conditions (soil type, fertility), the water and ecological system and the landscape typologies play an important role in deciding on a spatial strategy, which emphasizes the potentials for growing certain crops. The potentials can be natural, such as the availability of water, sunshine or the fertility of the soil, but they may also artificially defined, such as hydro- or aquaponics, which profit from microclimatic circumstances, such as the advantages of humid environments inside buildings, specific exposition to sunlight or shade at the lee side of buildings where wind doesn't get any grips. Besides natural forces, food potentials are also determined by economic factors, such as land value, the effectiveness of the chain and a market to sell to, or social aspects, such as the thrive in a neighbourhood to put effort in planting, harvesting and maintaining the space, social cohesion and the availability of active institutions. These food potentials determine what is possible and the most logical choice for a certain type of urban agriculture, but they do not decide on future use. Food has, especially in an urban context with many other spatial claims, to compete for space. There is one circumstance that might be helpful and force thinking in a supportive direction. We know from earlier research there is an estimated 30% space of the total area required to anticipate the impacts of climate change (Roggema 2013). When this space is created in urban regions to anticipate the impacts of climate change, such as flooding, droughts or fires, most of the time these spaces are unused. The climate hazard does not occur on a daily basis. This implies also these spaces can be used for other purposes for most of the time. Urban agriculture is a good choice, together with urban green space for leisure, parks or nature, as the characteristic of urban agriculture, being growing and re-growing living products, makes it easy to cope with an incidental disaster. The design strategy for a city region must then subsequently be oriented on the streamlining of both interests, linking the food-potentials with vulnerable areas. This way, the need for bigger public spaces for climate impacts and the desire to grow more local food can be connected end realised.

A design strategy helps to determine what to make and do, why to do ‘it’ (e.g. the localisation of urban agriculture) and how to innovate contextually, both immediately and over the long term (http://www.design-strategy.net/design-strategy). Such a strategy is useful for the larger scale and longer term, in a connected system of overarching goal setting and city-ambition.

Many schools have pioneered the thinking that has contributed to the theory and practice of design strategies, such as The Hochschule für Gestaltung Ulm (http://www.hfg-archiv.ulm.de/english/the_hfg_ulm/), with Tomás Maldonado (Lindinger 1991), the Centre for Design Management at London Business School (Gorb 1990) and Jay Doblin at IIT (http://www.aiga.org/medallist-jaydoblin/).

As an example, a design strategy formulates the ambition of a design and emphasizes what should be achieved on the long term. A design strategy could, when it is framed as a food strategy, determine the places in the city-region network to grow food.

1.5.2 The ‘In-Between’: Design Concept

The intermediate level connects the underlying landscape with local isolated spaces or projects. There is a systemic approach required, which forms an interconnection between water, energy, food, ecology, economic, mobility and social systems. The conceptual choices are important as they make the difference between successful connectivity and failure. Is the waste of one system beneficially used in another system? Do the isolated spaces receive the required supplies at the right moment? Is there enough capacity and flexibility in these systems to adjust at any given time? Are the systems designed with a conceptual eye-mark of supplying, storing and slow flowing? Do they have the capacity to hold and deliver to prevent downstream problems and provide resources upstream? A conceptual design approach is required to link the potentials of the underlying landscape, such as the existing soil quality, climatic conditions, availability of water, nutrients and energy, the value of the land or the social constructs, with the demands of the localised places in the urban environment.

These connecting systems operate at many different scales regulating in- and out-flux of goods and materials, such as waste for one system that is supplied to another system in another place. These systems guard the balance of demand and supply for many resources, with the objective to support individual places to grow and flourish. It can be the delivery of little bits of water to vegetables planted on a roof terrace or about the stream of the river providing the water and sediments that are necessary to supply an entire city or a region with fertility.

At this level conceptual designs provide the framework for further elaborative detailed designs and create the linkages between individual projects. Moreover, the conditions of the landscape, e.g. the potentials, determine which and how the systemic aspects are designed and form the basis for concrete designs. The urban systems link closely with the circular metabolism principles of closing cycles, recycling and reuse. At this level the existing city is analysed and the spatial opportunities to locate potential food-growing places identified. This leads to the definition of spatial typologies (Point to Point Communicatie 2013), such as place-making farming, rooftop farming and high-tech farming (all three located inside cities) or multifunctional farming or open field farming (both located at the urban fringes). Urban systems determine the localisation of these typologies, e.g. the place they can take up in the water-, nutrients-, transport- or energy flows. These systems also define the opportunities to find space for each of these types, based on the analysis of the existing city, and the development of specific design concepts.

'A design concept is an abstract idea in the form of a plan or intention, conceptualising a mental image which corresponds to a distinct entity or class of entities, or to its essential features, or determines the application of a term (especially a predicate), and thus plays a part in the use of reason or language' (http://www.oxforddictionaries.com/definition/english/concept). 'A design concept is an abstract representation of reality that acts both as a window to the problem situation as well as a window to solution possibilities. It is an empirical interpretation and understanding of what 'is' as well as a normative idea (conception) of what can 'become'. A design concept functions as a bridge between the problem view and the solution view and is the outcome of a search process that alternately takes a solution-driven and a problem-driven perspective. Problem and solution are inseparable in design tasks' (De Jonge 2009). A good example of such a design concept is CPULs (Viljoen 2005), defined as: 'Open landscapes productive in economical and sociological and environmental terms. They will be placed within an urban-scale landscape concept offering the host city a variety of lifestyle advantages and few, if any, unsustainable drawbacks'. A design concept is needed when the design strategy needs to be implemented in the city, and it functions as an abstract spatial image for the productive city.

1.5.3 The Top: Design Principles

Isolated spaces, no matter of what size they are, have as a result of their location, a specific potential for specific urban agriculture types. Through their linkages with the underlying landscape and the position in the integrated systems of water, energy, nutrients and biodiversity, these places all have a unique 'Urban-Ag DNA'. The choice of crops, the intensity of farming and the choice for a certain technological system are all determined by the position of this space in the system. The design principles must accommodate this specificity. Therefore, the isolated projects or spaces deserve the design principles that shape these places in a way they can operate within the systems, feeding and extracting resources to and from the project and to use and maintain the underlying landscape. The design of these places then takes place when it is actually establishing form.

At the upper level individual design projects are embedded in and related to both the urban systems level as the basic conditional underlying landscape. These projects, the 'inserted productive urban landscapes' (Viljoen 2005), must be linked with the other two levels, in order to prevent a footloose and stand-alone operation to grow food, which, most likely, are easy to remove when subsidy or volunteers are lacking. Therefore the design principles used need to attach with the specific genius loci of the place in the urban system. Questions such as 'What is their position in the water, ecological and nutrient system?', must be answered first, before design principles can be applied in an elaborated design.

A design principle is used when concrete designs must fit in the overarching design strategy and abstract design concept. Here the concrete spatial performance and appearance is executed along the lines of aesthetics and urban form, fitting in the strategy to realise a productive city and the abstract concept that links coherently the different parts of the productive city.

There are many design principles. These principles are used while designing and shaping the design. The way these principles are applied in the designs depends on the aim and context of the design task. The principle distinguishes the one design from the other. It creates harmony, or not. It creates balance, or not. And so on. The principles of landscape design include the elements of unity, scale, balance, simplicity, variety, emphasis, and sequence as they apply to line, form, texture, and colour. These principles are interconnected in the design (http://www.ext.colostate.edu/mg/gardennotes/413.html).

Unity (and Harmony) is the Quality of Oneness.

When all elements are in agreement, a design is considered unified. No individual part is viewed as more important than the whole design. A good balance between unity and variety must be established to avoid a chaotic or a lifeless design. To achieve visual unity is a main goal of design. When all elements are in agreement, a design is considered unified. No individual part is viewed as more important than the whole design. A good balance between unity and variety must be established to avoid a chaotic or a lifeless design (White 2011). This can be achieved using different methods (http://desktoppub.about.com/cs/basic/g/principles.htm):

- Proximity gives a sense of distance between elements or how close together or far apart items are to each other
- Similarity comprehends the ability to seem repeatable with other elements
- Continuation: the sense of having a line or pattern extend
- Repetition: elements being copied or mimicked numerous times—repetition use of the same colours, styles, shapes, or other elements and principles throughout a document

- Rhythm: is achieved when recurring position, size, colour, and use of a graphic element has a focal point interruption. Altering the basic theme achieves unity and helps keep interest
- Unity how well parts of the document work together (see proximity)
- Consistency uniform use of design elements.

Balance represents the distribution of elements.

It is a state of equalized tension and equilibrium, which may not always be calm. Formal balance repeats the same left and right, giving stability, stateliness, and dignity. It is high maintenance keep both side similar. Informal balance differs from left to right giving curiosity, movement, and feels alive. Total mass of plants need to balance left and right (http://www.ext.colostate.edu/mg/gardennotes/413.html). White (2011) distinguishes several types of balance http://desktoppub.about.com/cs/basic/g/principles.htm:

- Symmetry.
- Asymmetrical produces an informal balance that is attention attracting and dynamic.
- Radial balance is arranged around a central element. The elements placed in a radial balance seem to 'radiate' out from a central point in a circular fashion.
- Overall balance is a mosaic form of balance, which normally arises from too many elements being put on a page. Due to the lack of hierarchy and contrast, this form of balance can look noisy.

Simplicity and variety work together to balance each other.

Simplicity is the degree of repetition rather than constant change, which creates unity. Variety is diversity and contrast in form, texture, and colour preventing monotony (http://www.ext.colostate.edu/mg/gardennotes/413.html).

Hierarchy leads the user through each design element in the order of its significance. The type and images are expressed starting from the most important to the least. Hierarchy: A good design contains elements that lead the reader through each element in order of its significance. The type and images should be expressed starting from most important to the least (White 2011).

Scale/proportion uses the relative size of elements against each other and can attract attention to a focal point.

When elements are designed larger than life, scale is being used to show drama (White 2011). The following types are distinguished (http://www.ext.colostate.edu/mg/gardennotes/413.html):

- Absolute scale relates the comparative value of landscape elements to a fixed structure (for instance a house).
- Relative scale relates to comparative relative sizes or "values" of objects in the landscape. Relative scale is very emotionally charged and closely linked to colour. It may create a feeling of relaxation and peacefulness or one of energy and action.

- Low scale is relaxing and calming. It is used in the home landscape to give a feeling of peace and relaxation.
- High scale promotes action. It is used around large buildings and in large spaces to fill the space. Use of high scale in small spaces makes the space feel smaller.

Dominance/emphasis creates a focal point.

Emphasis is the dominance and subordination of elements (http://desktoppub.about.com/cs/basic/g/principles.htm). The human mind looks for dominance and subordination in life. As we look at a landscape from any direction, we need to see dominance and subordination of various elements. If we don't find it, we withdraw from the landscape. Some gardens lack the dominant element. Others suffer with too many dominant elements screaming to be the focal point. Emphasis can be achieved through different sizes, bold shapes, groupings, and the unusual or unexpected (http://www.ext.colostate.edu/mg/gardennotes/413.html). Dominance is created by contrasting size, positioning, colour, style, or shape. The focal point should dominate the design with scale and contrast without sacrificing the unity of the whole (White 2011).

Similarity and contrast make focal points visible through consistent design.

Too much similarity is boring but without similarity important elements will not exist. An image without contrast is uneventful so the key is to find the balance between similarity and contrast. Planning a consistent and similar design is an important aspect of a designers work to make their focal point visible (White 2011).

There are several ways to develop a similar environment (White 2011):

- Build a unique internal organization structure
- Manipulate shapes of images and text to correlate together
- Express continuity from page to page in publications. Items to watch include headers, themes, borders, and spaces
- Develop a style manual and stick with the format
- Similarly, contrasting environments can be developed in several ways
- Filled or empty space, space is near or far, and can be 2-D or 3-D
- Position to the left or right, in an isolated or grouped way, centred or off-centre
- The form can be simple or complex, beautiful or ugly, whole or broken
- The direction of the environment can be developed towards stability or movement
- The structure can be organized or chaotic, mechanical or hand-drawn
- The size can be large or small, deep or shallow, fat or thin.
- The colour is grey-scale or coloured, light or dark
- The texture can be fine or coarse, smooth or rough, sharp or dull
- The density can be transparent or opaque, thick or thin, liquid or solid
- The gravity can be light or heavy, stable or unstable.

Sequence is the change or flow in form, colour, texture, and size giving movement or life.

Movement is the path the viewer's eye takes through the artwork, often to focal areas. Such movement can be directed along lines edges, shape and colour within the artwork. Use proportionally larger numbers of fine textured elements. Texture becomes finer when distance increases. In a distant corner, finer textures are better, sequencing to coarser textures on the arms. There are few basic rules how many warm and cool colours to use. More is not always better. As a rule-of-thumb, designs need 90% green to set off 10% colour. Darkest shades and the purest intensity dominate and should be used at the focal point. Warm colours work best in sequence. Using cool colours in contrast is more effective than sequences (http://www.ext.colostate.edu/mg/gardennotes/413.html).

These design principles are applied to lines, shapes, textures and colour.

- A line connects and defines the space, and creates outdoor rooms. Lines are a powerful design element that define rooms and connect people to the landscape. For a professional touch, use sweeping bold lines and curves rather than small zigzags and small wavy curves (http://www.ext.colostate.edu/mg/gardennotes/413.html).
- Shapes or form include the three-dimensional mass. Line, direction, and arrangements determine the shape. The mass resulting from this influences the scale. For unity, the topographical characteristics can be repeated in the form of design elements. There are several types of form (http://www.ext.colostate.edu/mg/gardennotes/413.html).
- Horizontal and spreading forms emphasis the lateral extent and breath of space. These are comfortable shapes because it corresponds with the natural direction of eye movement.
- Rounded forms are most common in natural materials. They allow for easy eye movement and create a pleasant undulation that leads itself to certain groupings.
- Vase-shaped trees define a comfortable "people space" beneath the canopy.
- Weeping forms lead the eye back to the ground. What is below the weeping form often becomes a focal point.
- Pyramidal forms direct the eyes upward. Grouping pyramidal forms will soften the upward influence. They will look more natural in the surroundings with foliage to the ground.
- Texture is fine or coarse, it can be heavy and light, and thin or dense, emphasising light or shade. At a distance, texture comes from the entire mass effect of design elements and the qualities of light and shadows (http://www.ext.colostate.edu/mg/gardennotes/413.html).
- Colour gives greatest appeal, and evokes the greatest response. How does colour speak to you? Colour is powerful in creating mood and feeling. "Colour therapy" is a popular topic in our rapid paced modern world. What moods and feeling do various colour create for you? What colours work for the landscape? What moods and feeling do you want in the environment? Is it a space for relaxation and healing or a space for activities (http://www.ext.colostate.edu/mg/gardennotes/413.html).

1.6 Three Levels Connected

Each of the levels of the spatial sandwich are not representing a specific spatial scale, nor do design aspects such as design strategy, -concept and -principle. These can be applied at any given scale. However, once the strategy is set, the design-concept must alienate with the strategy and the use of design principles lives up to the expectation set in the strategy and concept. In Table 1.1 the different characteristics of the three levels of the spatial sandwich are presented.

The bottom level emphasises a strategic approach for long-term design interventions, which influence the whole landscape, for defining the potentials for productivity in combination with the strategic reservation for climatic impacts. These design strategies fit often in structure planning phases.

The level of urban systems, in the middle, deals with design at a conceptual, abstract level with a view on functionality and resilience of these systems. The combination of functionalities of these systems, integrated with each other, and the space determined for food production, leads to a spatial typology of urban agricultural design concepts. These concepts play an important role at a tactical planning level, in the phase of Master Planning.

The individual projects at the top level are important in an operational sense, where concrete project designs are required and specific design principles suit the elaboration of strategy and concept. Here concrete garden design and the design of public or private spaces are at hand, such as parks, rooftops, squares and private or public gardens.

The three levels of the spatial sandwich belong together. It is important to create coherence and allow mutual influences. Only then the projects can be located at the right places, creating an urban environment in which natural flows are most efficient and the city will be sustainable and productive. This coherence works in two ways: from the bottom level up to the upper and vice versa. Defining the strategy, conceptualised in a design concept may lead to specific designs for certain spaces, or a single initiative may start to develop at a certain location and influences the way urban systems are designed, adding or using material flows to the underlying landscape.

Table 1.1 Characteristics of three levels of the spatial sandwich

	Bottom	In-between	Top
Design type	Strategy	Concept	Principle
Mode of operation	Strategically	Tactically	Operationally
Focus of design	Entire landscape	Urban System	Individual project
Design aspects	– Food potentials – Space for climate	– Spatial typologies – Creating space in existing city	– (Garden) architecture – Park/public space design
Planning type/level	Structure plan	Master plan	Project design

1.7 Conclusion

The design of Urban Agriculture consists, as illustrated in this chapter of two sides: the demand side of how much, what and where agricultural products are needed and the design side, which aims to design attractive urban environments. These two sides are bridged, and where they come together the core question is: Which spaces can be found to create a productive city? Productivity of the city consists of the economic (earn money), social (come together), and food (supply agricultural products) aspects. Connecting these aspects with the design offerings at different levels (see Fig. 1.9), provides the level of though needed to enhance the productivity of the city.

Urban Agriculture is a relatively new phenomenon. As we have seen, the majority of food production has taken place outside the city boundaries. Whether this is a right or wrong assumption, it has been, for a long time, the case. Recently the opinion has changed. Urban Agriculture is an aspect of city life, which not only provides food, but also social innovation, spatial quality and more. Here it becomes a design task. The many individual projects around have shown that design of these isolated project areas is sometimes done with an eye on spatial dimensions, but often this isn't the case. People tend to discover: 'Hey, our project just started and we have grown step by step, without thinking of the way our project looks or what kind of overall design objective we had'. This emergent way of developing can't consider higher levels of scale and coherence, which can be used to improve the design quality at lower scales.

Therefore, there is a need for an integrated and broadened design approach, in which attention is paid to the connections between a design-strategy, -concept and -principles, at each project level (e.g. at any possible spatial scale). Within this broad approach systems thinking must be included, which connects and integrates water, energy, food, nutrients, biodiversity and climate systems with each other and with the project at hand.

The challenge ahead of us is to prove the contribution of these integrated and broad designs. In order to do so we need to further develop the design examples to showcase all the levels of the McAg, illustrating how individual projects gain coherence when these are based on higher-level strategic and conceptual design thinking.

References

Corbusier, L. (1935). *La Ville Radieuse*. Boulogne-sur-Seine: Editions de L'Archi-tecture de Aujourd'hui.

De Jong, T. M., & Van der Voort, D. J. M. (Eds.). (2005). *Ways to study and research urban, architectural and technical design*. Amsterdam: IOP Press BV.

De Jonge, J. M. (2009). *Landscape architecture between politics and science, an integrative perspective on landscape planning and design in the network society* (Ph.D. Thesis). Wageningen: Wageningen University and Research Centre.

Gorb, P. (1990). *Design management. Papers from the London Business School*. London: Architecture Design and Technology Press.
Gorgolewski, M., Komisar, J., & Nasr, J. (2011). *Carrot city, creating places for urban agriculture*. New York: The Mona-celli Press.
Hofstrand, D. (2011). *Can the world feed nine billion people in 2050? AgMRC Renewable Energy and Climate Change Newsletter*, November 2011. http://www.agmrc.org/renewable-energy/renewable-energy/can-the-world-feed-nine-billion-people-by-2050/. Accessed 7 November 2016.
Howard, E. (1902). *Garden cities of tomorrow*. London: S. Sonnenschein & Co., Ltd.
Lindinger, H. (1991). *Ulm design: The morality objects*. Cambridge: The MIT Press.
Miazzo, F., & Minkjan, M. (Eds.). (2013). *Farming the City. Food as a tool for today's urbanisation*. Haarlem: Trancity and Amsterdam: Valiz.
Philips, A. (2013) *Designing urban agriculture. A complete guide to the planning, design, construction, maintenance and management of edible landscapes*. New Jersey: Wiley.
Point to Point Communicatie (Red.). (2013). *Stadsboeren in Nederland; Professionalisering van de Stadsgerichte Landbouw*. Rotterdam: Ministerie van EZ, Ministerie van IenM, Van Bergen Kolpa Architecten, LEI, De Volharding Breda, Priva.
Roggema, R. (2013). *Swarm planning: The development of a methodology to deal with climate adaptation*. Springer Theses, Recognizing Outstanding Ph.D. Research. Dordrecht, Heidelberg, London: Springer, 286 pp.
Roggema, R. (2014a). Framing urban agriculture: The quest for new design concepts. In *Proceedings ECLAS-conference*, 21–23 September 2014, Porto.
Roggema, R. (2014b). It's time for the McAg: Finding spaces for productive cities in a spatial sandwich. In *Proceedings IFLA 2014*, 4–6 June 2014, Buenos Aires.
Sijmons, D. F. (1992). Het Casco-concept; Een Benaderingswijze voor de Landschapsplanning. Studiereeks Bouwen aan een Levend Landscap, nr. 24. Utrecht: Directie Bos en Land schapsbouw, Ministerie van Landbouw, Natuurbeheer en Visserij.
Steele, C. (2008). *The Hungry City, How food shapes our lives*. London: Chatto and Windus.
Tjallingii, S. P. (1993). *Ecopolis: Strategies for ecologically sound urban development*. Leiden: Backhuys Publishers.
Van den Dobbelsteen A., Jansen, S., van Timmeren, A., & Roggema R. (2007). Energy Potential Mapping—A systematic approach to sustainable regional planning based on climate change, local potentials and exergy. In *Proceedings of the CIB World Building Congress 2007*. Cape Town: CIB/ CSIR.
Viljoen, A. (Ed.). (2005). *CPULs: Continuous Productive Urban Landscapes. Designing urban agriculture for sustainable cities*. Amsterdam, Boston, Heidelberg, London, New York, Oxford, Paris, San Diego, San Francisco, Singapore, Sydney, Tokyo: Architectural Press, Elsevier Ltd.
Von Thünen, J. H. (1826). *Die isolierte Staat in Beziehung auf Landwirtshaft und Nationalökonomie*. Pergamon Press, New York. (English translation by Wartenberg C. M. in 1966, P.G. Hall, editor).
White, A. (2011). *The elements of graphic design*. New York: Allworth Press. ISBN 978-1-58115-762-8.
Wright, F. L. (1945). *When democracy builds*. Chicago: University of Chicago Press.

Websites

http://www.aiga.org/medalist-jaydoblin/. Accessed 29 May 2014.
http://boundariescrossing.wordpress.com/2012/07/17/research-through-design/. Accessed 25 March 2014.
http://www.design-strategy.net/design-strategy. Accessed 10 February 2014.
http://desktoppub.about.com/cs/basic/g/principles.htm. Accessed 21 February 2014.

http://www.ext.colostate.edu/mg/gardennotes/413.html. Accessed 25 February 2014.
http://www.hfg-archiv.ulm.de/english/the_hfg_ulm/. Accessed 29 May 2014.
http://www.mvrdv.nl/projects/181_pig_city/#. Accessed 27 April 2014.
http://www.oxforddictionaries.com/definition/english/concept. Accessed 10 February 2014.
http://www.oxforddictionaries.com/definition/english/design. Accessed 10 February 2014.
http://www.oxforddictionaries.com/definition/english/planning. Accessed 10 February 2014.

Chapter 2
Space for Food in the City

Rob Roggema

2.1 Introduction

An analysis of the total food consumption in the Netherlands and the estimated production of food within urban boundaries demonstrate that only 0.0018% of the consumption is currently produced in the city (Roggema 2015). In order to increase this percentage the city needs to offer more space (Roggema 2014a, b, c). The question is where to find space in urban environments, which often suffer from land use competition and an urge for higher densities. There are three basic options: lowering the density, transformation of existing uses, and find space under and above buildings (Roggema 2015). Often these options are underused, partly because a spatial analysis of the linkages between available space and urban agriculture types lacks. In this chapter the potential link between capacity and spatial typology in the city is investigated, resulting in a potential increase of urban production.

For long, urban design has not included Urban Agriculture (see Chap. 1). The majority of current food-projects in the city is small, is often located on temporary, unused space, and developed against the odds by persisting residents. An analysis at city level is often lacking. This leads to an underuse of suitable urban spaces, and the occurrence of misfits, for instance when a project has larger demands than seem to be available. But if we take a look at the entire city the capacity is larger than

R. Roggema (✉)
Faculty of Design, Architecture and Building, University of Technology Sydney, Ultimo, NSW, Australia
e-mail: rob@cittaideale.eu

R. Roggema (ed.), *Food Roofs of Rio de Janeiro*,
DOI 10.1007/978-3-319-56739-6_2

expected. Recent analysis shows for instance that in Amsterdam 12% of the area could be transformed into productive spaces, providing food for approximately 25% of the Amsterdam population (Mulder and Oude Aarninkhof 2014). In this calculation valuable space, rooftops and private land is excluded. A big win could be achieved when the right fit between available space (capacity) and the type of urban agriculture (typology) is made.

2.2 Methodology

The method presented in this chapter consisted of five steps. First, desktop research and analyses of urban agriculture designs led to a collection of good and diverse examples. Secondly, definitions of spatial capacity and typology have been developed. In the third step explorations on spatial dimensions of the capacity and typology was undertaken in order to estimate average sizes of urban spaces and urban agriculture types. In the next step the different elements of typology and capacity are combined in a matrix. When subsequently capacities and typologies are maximally connected at city level an estimation of the productivity is made and compared with current productivity of food production in the city.

2.3 Definitions

In order to identify space in the city for urban food production the capacity and typology are matched. The following definitions of spatial capacity, Urban Agriculture typology and Capacity-Typology Matrix have been formulated:

1. The Spatial Capacity is defined as the amount of public and private spaces, which potentially could be made available as productive areas.
2. Urban Agriculture Typology is defined as the different spatial manifestations of Urban Agricultural types in the urban and peri-urban areas with their spatial specifications.
3. The Capacity-Typology matrix is defined as a table, which offers insights in the possible matching of urban spaces with Urban Agricultural types.

2.4 Explorations on Spatial Capacity and Typology

Given the idea many people have that it is impossible to produce enough food in urban environments to feed the urban population, it is necessary to understand which areas in the city are potentially available to grow food, which types of Urban Agriculture might fit with these available spaces and how much food could be grown at an urban level. Therefore the first element to investigate is the available and potentially usable Spatial Capacity in the city. The second step is to identify the types of Urban Agriculture and define their spatial dimensions. Once the Capacities and Typologies are known they can be matched in a matrix. An estimate of the number of types that could appear in an average city will subsequently lead to a total amount of food, which potentially can be produced in the city.

2.4.1 *Spatial Capacity*

The capacity of the city is measured in terms of available spaces for growing food. Firstly, a range of spaces, both public, such as parks, nature, rest spaces, infrastructure, parking (Table 2.1), and private, such as roofs, balconies, gardens, parking, facades, inside buildings (Table 2.2) have been identified, their sizes are

Table 2.1 Private spaces and their dimensions

Private space	Specifications	Size	Use	Conditions
Roofs	Family	10 × 15 m	PV, empty	Construction, combination with PV
	Company	30 × 50 m	PV, empty	Construction, combination with PV
	Collective	20 × 30 m	PV, empty	Construction, combination with PV
Balconies	Family	2 × 3 m	Storage, laundry, bbq's	Pots, small
Facades	Family	10 × 6 m	Unused	Construction
	Company	40 × 40 m	Unused	Construction
Gardens	Family	5 × 5 m	Concrete, flowers, gardening, containers	Possible
	Company	30 × 30 m	Water, meadow, trees, for looks	Possible
Parking	Family	4 × 15 m	Partially for P	Use only unused P's
	Company	20 × 20 m	Partially for P	Use only unused P's
Building	Home	10 × 15	Living	Space inside
	Flat/apartment	15 × 40 m	Living	Spaces inside
	Industrial/warehouse	50 × 20 m	Manufacturing	Reuse of internal spaces

Table 2.2 Public spaces and their dimensions

Public space	Specifications	Size	Use	Conditions
Park	Big	100 × 100 m or more	Recreation, nature, dogs, mix	Only partly use, leisure is core
	Small	30 × 40 m	Recreation, play	Leisure is prime
Public green	In between houses/buildings	50 × 20 m	Playground, field-games (football), dogs	Use the underused only
Nature	Eco-zone	50 × 50 m	Nature	Ecological quality is main factor
	Connections	15 m × length	Nature, leisure, mobility	Ecological connections must prevail
Rest spaces, left-overs	Corners, mis-designs	5 × 10 m	–	–
Along infra	Berms	15 m × length	–	Pollution must be solved
	Wedged space	20 × 40 m	Nature, unused	Possible pollution should be solved
Parking	Collective	50 × 50	Parking	Only underused parking area, temporarily used; double use
	In streets	3 × 5 m	Parking	Transform from parking into productive

estimated and the possible combinations of use are determined. When in a city all these spaces are located a map it could make the capacity for growing food in the city visible.

2.4.2 *Urban Agriculture Typology*

A range of types, with according sizes, requirements, types of use and examples have been found. These types are subdivided in productive spaces (urban farm, aquaponic system, roof garden, collective garden, edible forest garden, edible forest, wild foraging, fringe farm, vertical farm, private vegetable garden, allotment garden, productive garden/restaurant, productive garden/shop, productive garden/hotel and productive garden/office), distribution (individual, company and trucks), processing (factory, at home) and sale (street vendors, restaurant, shop, market). The properties of each type is characterised in Table 2.3. The specification, approximate size, the use (for own use or for sale), what kinds of combinations in use are possible what kind of requirements and some examples are given. Each of these typologies requires space, but in a different manner.

Table 2.3 Properties of urban agriculture typologies

Type	Specification	Approx. size	Own use/sale	Use/mix	Requirements	Example
	Productive space					
Urban farm		100 × 50 m	Sale	Building, animals, outside and inside, pigs, chicken	Building, space	Caetshage, EVA-Lanxmeer, NL[1]
Aquaponic system	Roof	10 × 5 m	Own use	Leisure, wash, PV	Construction, water, fish	FoodRoof Rio, Brazil[2] (Roggema 2014c; Roggema et al. 2014)
	In building	20 × 30 m	Own use/sale	Visitor centre	Construc-tion, water, fish	Biospheric project Manchester, UK (Keeffe 2015; Roggema 2014b)[3]
Roofgarden	House	10 × 15 m	Own use	Terrace, PV	Construction, maintenance	FoodRoof Rio, Brazil
	Flat/office	15 × 50 m	Own use	Lunch break, PV	Construction, maintenance	Zuidpark Amsterdam, NL (Mulder en Oude Aarninkhof 2010)
						Ebisu metro-station, Tokyo, Japan[4, 5] (Xie undated; Plaskoff Horton 2014)
Collective neighbourhood vegetable garden		20 × 30 m	Own use/sale	Education, social cohesion	Volunteers	Parmedines, Wijsgerenbuurt, Geuzenveld, Amsterdam, NL[6]
Edible forest garden		20 × 20	Own use/sale	Leisure, social cohesion	Space	Sarphatipark, Amsterdam, NL[7]
Edible forest		100 × 50	Sale	Recreation, nature, social cohesion	Forest, space	Voedselbos Vlaardingen, NL[8]
						Voedselbos Makeblijde, Houten, NL[9]
Wild foraging		?	Own use	Recreation, experience, urban activities	Unexpected food in urban places	Urban edibles, Portland, Oregon, US[10]
						BOSKOI, urban edibles app, NL[11]
Fringe farm		200 × 150 m	Sale	Care, nature, education, meeting facility	Space, manager	Hoge Born, Wageningen, NL[12] Kemphaan, Almere, NL[13]

(continued)

Table 2.3 (continued)

Type	Specification	Approx. size	Own use/sale	Use/mix	Requirements	Example
Vertical farm	House	10 × 10 m façade	Own use	Living	Owners, maintenance, harvest	
	Whole building complex	40 × 20 m façade	Sale	Office	Owners, maintenance, harvest	Green spirit farms new buffalo, Michigan, US[14]
						Plantagon, Sweden[15]
						The Living Skyscraper, Chicago, Blake Kurasek US[16]
Vegetable garden	Private	10 × 10 m	Own use	Flowers, meadow, shed	Residents	
Allotment gardens		100 × 100 m	Own use	Townhouse, flowers, nature, collective shed	Association	Nut en Genoegen, Sloterdijk, Amsterdam, NL[17]
Productive garden annex	Restaurant garden	30 × 20 m	Sale		Grow the demand	Uit je eigen stad, Rotterdam, NL[18]
	Shop garden	10 × 25 m	Sale		Clients	Hemel, food garden farm shop, Hemel Hempstead, UK[19]
	Hotel garden	10 × 15 m	Sale		Restaurant kitchen	Brubacher Homestead Bed and Breakfast, Elmira, Canada[20]
	Office garden	20 × 30 m	Own use/sale		Employees	LED vegetable garden, Yokohama, Japan[21]
	Distribution					
Individuals	Garage	10 × 5 m	Sale	Living	Storage, transport, (electric bike), web	Web-shop Esther, Wageningen, NL
	Bike-paths	2 m × ?				
Companies		30 × 50	Sale		Employees, web-shop, transport/small trucks shack	Hello Fresh, international[22]
	Roads:	10 × ?m				
Trucks	Shack roads	50 × 50, 10 × ?m	Sale		Chauffeurs, web-shop, roads, shack	Refinery29, New York, US[23]

(continued)

Table 2.3 (continued)

Type	Specification	Approx. size	Own use/sale	Use/mix	Requirements	Example
	Processing					
Factory		50 × 50 m	Sale	Shop	Equipment, employees Close to production and client	Kicking horse coffee company and Kootenay Pasta, Columbia Valley, Canada[24]
						De Witte Lelie, Maasland, NL[25]
						Titi Eco Farm, Malaysia[26]
Home-made		10 × 10 m	Sale	Living	Equipment,	
					Distribution	
					Channel	
					Close to	
					Production and	
					Marketing	
	Sale					
Streetfood-vendors		5 × 4 m	Sale	Festival, leisure	Spacelocation. Close to supply clients	Rollende keukens, Amsterdam, NL[27]
Restaurant		10 × 20 m	Sale	Shop	Space and location, close to production and clients	Hooi, Utrecht, NL[28]
Shop		10 × 10 m	Sale		Supply, close to clients	Unicorn Grocery, Manchester, UK[29]
					Space/location	
Market		100 × 50 m	Sale		Distribution, close to clients	Whole foods Market, Fulham, UK[30]

[1] http://www.urbangreenbluegrids.com/projects/caetshage-urban-farm-eva-lanxmeer-culemborg-the-netherlands/
[2] https://nl-nl.facebook.com/pages/Food-Roof-Rio/301321946708071
[3]http://www.biosphericfoundation.com
[4]http://garakami.com/20140411/tokyo-train-station-opens-rooftop-garden-to-community/

[5] http://www.urbangardensweb.com/2014/03/27/japanese-commuters-can-tending-train-station-rooftop-gardens/
[6] http://www.buurtmoestuin.nl/index.php?option=com_content&view=article&id=17&Itemid=18
[7] http://www.visionair.nl/politiek-en-maatschappij/nederland/de-eetbare-bostuin/
[8] http://www.rfgn.nl, http://www.voedselbosvlaardingen.nl/p/planning.html
[9] http://www.foodforestry.nl/#!vbamb/cv7e
[10] http://urbanedibles.org
[11] http://boskoi.org/main
[12] http://hogeborn.lievegoed-bedrijven.nl
[13] http://www.kemphaan.nl
[14] http://www.greenspiritfarms.com/#!michigan/chuy
[15] http://plantagon.com
[16] http://blakekurasek.com/thelivingskyscraper.html
[17] http://www.nutengenoegen.magix.net
[18] http://www.uitjeeigenstad.nl
[19] http://www.sunnysideruraltrust.org.uk/hemel-food-garden/
[20] http://www.bbcanada.com/14037.html
[21] https://www.jetro.go.jp/en/ind_tourism/keystone.html
[22] https://www.hellofresh.nl
[23] http://www.refinery29.com/lunch-calling
[24] https://www.kickinghorsecoffee.com/files/2013_CV_Business.pdf
[25] http://www.netwerkplatteland.nl/groeten-uit-maasland-de-jamfabriek/
[26] http://www.titieco.com.my/index.php?option=com_content&view=category&layout=blog&id=47&Itemid=131
[27] http://rollendekeukens.amsterdam
[28] http://www.hooiutrecht.nl
[29] http://www.unicorn-grocery.coop
[30] http://www.wholefoodsmarket.com/stores/list/uk

2.4.3 Capacity-Typology Matrix

When the types of Urban Agriculture and the spatial capacities are brought together in a matrix, the required sizes of urban agriculture types can be matched with the available spaces. This matrix gives insights on where the different types of Urban Agriculture could fit best. In practice, and for each city specifically, the exact locations should be identified in detail, and further designed (Table 2.4).

2.4.4 Calculations on Capacities

Assuming that each of the Urban Agriculture types finds its fit in every municipality in the Netherlands it is possible to calculate the total amount of food, which can be produced within urban boundaries. For each type an average number that reasonably can be realised in a municipality is multiplied with the size in m^2 of the typology. The total area is subsequently multiplied by 661 kg, which is the average productivity per hectare in urban areas. In order to find the total potential amount of food produced within urban boundaries, this number is multiplied by 401, which is the number of municipalities in the Netherlands. In total 25.3 million kg's can be produced in urban environments (see Table 2.5).

To find out whether this is substantially contributing to the food supply in the Netherlands, the total amount of food produced within urban boundaries needs to be compared with the total amount of food consumed in the Netherlands. The food consumed in the Netherlands consists of meat and fish (Table 2.6), vegetables (Table 2.7) and drinks (Table 2.8).

The total consumption in the Netherlands is 13.552.237.838 (13.5 billion) kilogram. The 25.3 million kilograms of food produced within urban boundaries is approximately 18.7%. In order to raise this percentage it is necessary to extend the urban boundaries beyond the strict urban.

2.5 Elaborations at City Region Size

How much can we extend the boundaries of the urban conglomerate for local food production? What seems to be a reasonable assumption, is to double the current radius of a city. The total area of a city is πr^2. If we take the radius of this city and extend the city region all around it with a similar radius, the area of the city region is $\pi(2r)^2$, which is four times the area of the existing city. If we project this ratio to Amsterdam (Fig. 2.1), a city-region emerges, within which the majority of food production, distribution, processing and consumption could take place. In this image the average radius from the city centre (Dam square) to the current urban boundary is around six km. The city region is defined by extending urban boundary with the same distance.

Table 2.4 Capacity-typology matrix

Capacity type	Big park	Small park	Public green in between houses	Eco-zone	Eco-connection	Left-over corners	Infra-berms	Infra wedges	Collective parking	Street parking	Family parking	Company parking	Family roofs	Company roofs	Collective roofs	Balconies	Family façade	Company façade	Family garden	Company garden	In home	In apartments/offices	In warehouse
Production																							
Urban farm	■			■					■														
Aquaponic system, Roof													■	■	■	■	■	■					
Aquaponics, in building																					■	■	■
Roof-garden, house													■										
Roof-garden, flat/office														■	■								
Collective neighbourhood vegetable garden	■		■	■				■	■						■					■			
Edible forest garden	■	■	■	■	■			■	■			■											
Edible forest	■			■																			
Wild foraging	■	■	■	■	■	■	■	■	■	■		■											
Fringe farm				■																			
Vertical farm, house													■			■	■				■		
Vertical farm, building complex														■	■			■				■	■
Vegetable garden, private					■	■	■	■		■	■	■	■			■	■		■				
Allotment gardens	■	■	■	■				■	■			■		■	■					■			
Productive garden annex, restaurant			■						■						■			■		■			
Productive garden annex, shop			■					■	■			■		■	■			■		■			■
Productive garden annex, hotel	■		■					■	■			■		■	■			■		■			■
Productive garden annex, office									■			■		■	■			■		■		■	■
Distribution																							
Individuals		■	■			■		■				■	■	■	■	■	■	■	■	■	■	■	■
Companies	■		■					■	■			■		■	■			■		■		■	■
Trucks							■	■	■			■											
Processing																							
Factory	■		■					■	■			■		■	■								■
Home-made		■	■			■	■	■			■		■		■						■	■	
Sale																							
Streetfood-vendors	■	■	■			■		■	■	■		■								■		■	■
Restaurant	■	■	■	■				■	■			■		■	■					■			■
Shop	■	■	■			■		■	■			■		■	■								■
Market	■		■					■	■			■		■	■					■		■	■

Table 2.5 Total capacity of urban agriculture in the Netherlands

Type	Approx. size (m)	Size (m^2)	Number	Total area (m^2)	Productivity 10.000 = 661 kg[a]
Production					
Urban farm	100 × 50	5000	5	25,000	1652.5
Aquaponic system, Roof	10 × 5	50	80	4000	264.4
Aquaponics, in building	20 × 30	600	15	9000	594.9
Roof-garden, house	10 × 15	150	250	37,500	2478.75
Roof-garden, flat/office	15 × 50	750	150	112,500	7436.25
Collective neighbour-hood vegetable garden	20 × 30	600	80	48,000	3172.8
Edible forest garden	20 × 20	400	15	6000	396.6
Edible forest	100 × 50	5000	5	25,000	1652.5
Wild foraging	?	1000	5	5000	330.5
Fringe farm	200 × 150	30000	10	300,000	19830
Vertical farm, house	10 × 10 façade	100	75	7500	495.75
Vertical farm, building complex	40 × 20 façade	800	125	100,000	6610
Vegetable garden, private	10 × 10	100	1500	150,000	9915
Allotment gardens	100 × 100	10000	10	100,000	6610
Productive garden annex, restaurant	30 × 20	600	15	9000	594.9
Productive garden annex, shop	10 × 25	250	25	6250	413.125
Productive garden annex, hotel	10 × 15	150	10	1500	49.15
Productive garden annex, officeis	20 × 30	600	20	12,000	793.2
Total: 63,290.325 kg per municipality					
403 Municipalities × Total = 25,379,420.3 kg					
Total NL = 13,552,237,838 kg[b]			Urban Agriculture part = 18.7% Approx. 30 times current production		

[a]Based on a literature study of the following sources: De Graaf (2011a, b), De Muynck (2011), Dijksma (2013, 2014), Ecovrede (2012), Stadslandbouw (2009), Rotterdam (2012), Gorgelewski et al. (2011), Groene Ruimte (2014), Jansma et al. (2011), Kuypers (2012), Ladner (2011), Marsden and Morley (2014), Miazzo and Minkjan (2012), Philips (2013), Point to Point Communicatie (2013), Stadslandbouw (2010), Stutterheim (2013), Van der Sande (2012), Veen et al. (2012), Viljoen (2005)

[b]Based on a combination of data retrieved from: CBS (2009, 2014), Geurts et al. (2014), Goeievraag (2011), Nederlandse Vegetariërs Bond (2014), Productschap Vee en Vlees en het Productschap Pluimvee en Eieren (2011), Van der Bie et al. (2012), Van Rossum et al. (2011), Van Rossum en Geurts (2013), Voedingscentrum (2014)

Table 2.6 Consumption of meat and fish in the Netherlands on a yearly basis

	Kg/year	#/year
Cow	323,500,000	1,400,000
Pig	684,200,000	1,300,000
Chicken	353,800,000	283,000,000
Turkey-duck-goose	26,900,000	
Sheep-goat-horse	40,400,000	
Fish	538,000,000	
Total	1,966,800,000	

Sources Productschap Vee en Vlees en het Productschap Pluimvee en Eieren (2011); http://www.goeievraag.nl/vraag/maatschappij/samenleving/dieren-koeien-varkens-schapen-dergelijke.124647

Table 2.7 Consumption of vegetables in the Netherlands on a yearly basis

	Kg/year	#40 ft. containers
Potatoes	416,927,001	15,160
Legumes	557,009,223	20,254
Milk products	426,144,378	15,496
Cereals	1,630,701,926	59,298
Cakes	175,493,797	
Sugar	178,660,331	6496
Fat	117,295,530	4265
Total	3,502,232,186	

Source http://statline.cbs.nl/StatWeb/selection/?VW=T&DM=SLNL&PA=37296ned&D1=1-2,8-13&D2=0,10,20,30,40,50,(l-1)-l&HDR=G1&STB=T

Table 2.8 Consumption of drinks in the Netherlands on a yearly basis

	l/year	#40 ft container
Non alcoholic	7,692,067,983	295,848
Alcoholic	275,478,457	10,595
Condiments	115,659,215	4448
Total	8,083,205,652	

Source http://statline.cbs.nl/StatWeb/selection/?VW=T&DM=SLNL&PA=37296ned&D1=1-2,8-13&D2=0,10,20,30,40,50,(l-1)-l&HDR=G1&STB=T

Within the urban contour an estimated 12% (Mulder and Oude Aarninkhof 2014) could be used for urban farming, which means that 88% is in use for other purposes. In the surroundings of the city the percentage, which could be used for urban farming could be much higher, because the pressure of other uses is much lower. Instead of 12% we estimate that in the countryside the opposite percentage of 88% could be used for urban farming. If we look at the City Region scale, 88% of

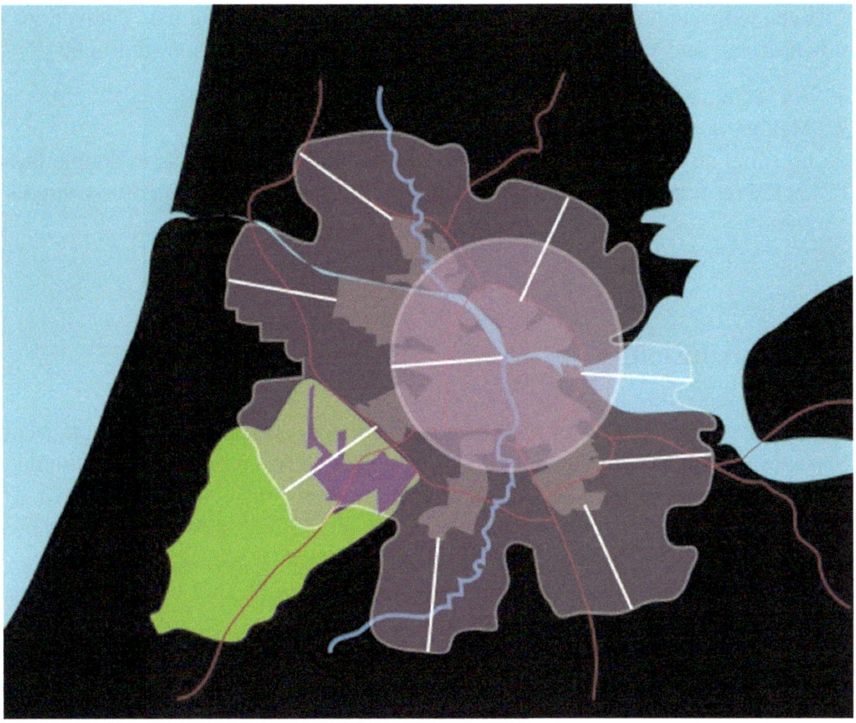

Fig. 2.1 Projection of the city-region Amsterdam

an area four times as big as the surface area of the current urban contour could be used for Urban Agriculture. If we take this example as the average for every municipality in the Netherlands we can calculate which percentage of the total consumption could be produced here. If 18.7% of the entire whole food consumption in the Netherlands can be grown within the urban contour (i.e. 12% of the whole surface area of the city), and 88% of the area outside the city boundaries, which is four times bigger than the city, can be used for urban farming, a total of 545% of the total food consumption can be produced within the City Region boundaries (i.e. 4 × 131.6% (88% of four areas as big as the city; if 18.7% fits on 12% of the city, 131.6% fits on 88% of the same area) + the original 18.7%). Hence, in and directly around the city, we could produce more than five times the amount of food consumed by the population.

These estimates are averages for big and small cities and/or municipalities. The larger cities, such as Amsterdam may perform lower percentages, because in these cities the population density is relatively high, but at the national level this will be compensated by the performance in lower density towns.

Even with the huge potentials urban and peri-urban farming has, innovations in urban design and technology could even increase the production, through:

1. An increase of the productivity per hectare;
2. Making additional space available to become productive;
3. Opening up extra space found in unexpected locations, such as in double layers on the roof, multiple floors under the building or transformation of existing land uses.

2.6 Conclusions

The calculations and sizes of specific uses and spaces are based on assumptions and averages. To get a clear picture of the exact potentials, every municipality should be investigated their own situation, or co-research this in a regional context. The calculations on capacity and typology of urban agriculture show there is much more space available to grow food in the city than is currently used. If all the capacity, in public and private spaces, are taken into account the potential of food production within urban boundaries could provide 18.7% of the Dutch national consumption, which is a huge leap forward compared with the currently produced amount (0.002%). To even increase this number the direct surroundings of the urban conglomerate come into play. Rough estimates on the potential productivity in the City-Region, which is set at being four times the size of the urban area, show the total production could be more than five times the demand. Because the calculations have been undertaken with quite some assumptions it would be good to investigate which exact area is required to feed the urban population.

New urban design concepts need however to be developed to further increase the space suitable to grow food, but also to make transformations from current uses insightful; and acceptable.

The FoodRoofs are one innovative addition to the spectrum of urban spaces and could therefore play a significant role in increasing the productivity of the city. Especially because the FoodRoof is located in possibly the hardest areas of the city and the potential space on the roofs in these areas is enormous. The FoodRoof in its precinct context is a concept that binds the water, nutrient and energy system together and therefore requires also design at the urban design level. It could well be that the traditional way of building cities is to be replaced by new and innovative ways of designing. Instead of increasing densities, a productive city has to create extra space to grow food. This search should focus on the transformation of existing land use, it should fully profit from the available green spaces and it should find unprecedented spaces in places that didn't qualify as productive spaces before.

References

CBS. (2009). *Nederlander eet iets meer kippenvlees*. Den Haag: Centraal Bureau voor de Statistiek. Re-trieved July 24, 2014 from: http://www.cbs.nl/nl-NL/menu/themas/industrieenergie/publicaties/artikelen/archief/2009/2009-11-24-vleesconsumptie-tk8.htm

CBS (2014). *Beroepsbevolking; kerncijfers naar ge-slacht en andere kenmerken 1996–2014*. Retrieved July 24, 2014 from: http://statline.cbs.nl/StatWeb/publi-cation/default.aspx?DM=SLNL&PA=71958NED&D1=5-11&D2=0&D3=0&D4=26%2c31%2c36%2c41%2c46%2c51&HDR=G3%2cT&STB=G1%2cG2&VW=T

De Graaf, P. (2011a). *Ruimte voor Stadslandbouw. Eetbaar Rotterdam in kaart. Lay-out 18*. Rotterdam: Stimuleringsfonds voor Architectuur.

De Graaf, P. (2011b). *Ruimte voor Stadslandbouw in Rotterdam*. Rotterdam: Stimulerinsgfonds voor Archi-tectuur, Vereniging Eetbaar Rotterdam.

De Muynck, A. (2011). *Stadslandbouw en Duurzame Gebiedsontwikkeling. Een onderzoek naar de bijdrage van Stadslandbouw aan duurzame gebiedsontwik-keling*.

Dijksma, S. (2013). *Kamerbrief Verkenning Stadsland-bouw*. Den Haag: Ministerie van EZ.

Dijksma, S. (2014). *Toespraak van Staatssecretaris Dijksma (EZ) voor de Dag van de Stadslandbouw in Utrecht op 15 mei 2014*. Den Haag: Ministerie van EZ.

Ecovrede. (2012). *Criteria voor Succesvolle Stadsland-bouw*. Brief aan de Gemeenteraad Arnhem. Rotterdam: Erasmus Universiteit Rotterdam, Stadsontwikkeling Rotterdam, Technische Universi-teit Delft.

Geurts, M., C. T. M. van Rossum, H. Brants, J. Ver-kaik-Kloosterman, & S. Westenbrink (2014). *Veran-deringen in het aanbod van voedingsmiddelen en de voedselconsumptie*: *Resultaten gebaseerd op bijna 25 jaar voedselconsumptieonderzoek*. Bilthoven: RIVM Rapport 090429001.

Goeievraag (2011). *Hoeveel dieren* (koeien/varkens/ schapen en dergelijke) eten wij als nederlanders per dag. Retrieved July 25, 2014 from: http://wwwgoeievraag.nl/vraag/maatschappij/samenleving/di-eren-koeien-varkens-schapen-dergelijke.124647

Gorgelewski, M., Komisar, J., & Nasr, J. (2011). *Car-rot City*. Creating places for urban agriculture. New York: Monicelli Press, Random House Inc.

Groene Ruimte. (2014) *Dossier Stadslandbouw*. Retrieved July 25, 2014 from: http://www.groen-eruimte.nl/dossiers/stadslandbouw/

Horton, R. P. (2014) *Japanese commuters tending train station rooftop gardens*. Published online, March 27, 2014 http://www.urbangardensweb.com/2014/03/27/japanese-commuters-can-tending-train-station-rooftop-gardens/

Jansma, J. E., Veen, E., Dekking, A., Vijn, M., Sukkel, W., Schoutsen, M., et al. (2011). *Staalkaarten Stadslandbouw + Ontwik-kelstrategieën om te komen tot Stadslandbouw in Almere Oosterwold*. PPO: Lelystad.

Keeffe, G. (2015). Why urban food production makes sense and how to make it happen. In: IUFN, published online January 29, 2015. http://www.iufn.org/en/iufninterview/keeffe/

Kuypers, V. (2012). *Stadslandbouw leuk*, maar niet meer dan dat. Binnenlands Bestuur.

Ladner, P. (2011). *The urban food revolution. Changing the way we feed cities*. Gabriola Island: New Society Publishers.

Marsden, T., & Morley, A. (Eds.). (2014). *Sustainable food systems. Building a new paradigm*. London and New York: Routledge.

Miazzo, F. & M. Minkjan (Eds.). (2012). *Farming the city. Food as a tool for today's urbanisation*. Amsterdam: Trancity*Valiz.

Mulder, M., & Oude Aarninkhof C. (2010). Productief Stedelijk Landschap. *Stedebouw and ruimtelijke ordening*, **(5)**, 40–43.

Mulder, M., & Oude Aarnibnkhof, C. (2014). *Stadslandbouwdoos*. Den Haag: Atelier Rijksbouwmeester.

Nederlandse Vegetariers Bond (2014). *Hoeveel dieren spaart een vegetarier in zijn hele leven uit?*. Retrieved July 24, 2014 from: http://www.vegetari-ers.nl/bewust/waarom-vegetarisch/voor-de-dieren/aantal-gegeten-dieren

Philips, A. (2013). *Designing urban agriculture. A complete guide to the planning, design, construc-tion, maintenance and management of edible land-scapes*. Hoboken: John Wiley & Sons, Inc.

Point to Point Pommunicatie (Eds.). (2013). *Stadsbo-eren in Nederland. Professionalisering van de Stads-gerichte Landbouw*. Rotterdam: Van Bergen Kolpa, LEI, De Volharding Breda, Paul de Graaf Ontwerp en Onderzoek, Priva, Ministerie van EZ, Ministerie van IenM.

Productschap Vee en Vlees en het Productschap Pluimvee en Eieren (2011). *Vee, Vlees en Eieren in Nederland, Kengetallen 2010*. Zoetermeer: PVV & PPE.

Roggema, R. (2015). The reinvention of the academic conference: How active delegates develop productive city concepts. *Future of Food: Journal on Food, Agriculture and Society, 3*(1), 63–78.

Roggema, R. (2014a). *Finding spaces for productive cities*. Keynote lecture, 6th AESOP conference sustainable food planning. Leeuwarden. November 6, 2014.

Roggema, R. (Ed.). (2014b). *Greg keeffe: The nutritious city. The biospheric foundation* (p. 51). International urban agriculture lecture, March 5, 2014. Velp: VHL University of Applied Sciences. ISBN: 978-90-822451-0-3.

Roggema, R. (2014c). FoodRoof Rio: How Favela Residents Grow Their Own Food. *Adjacent Government, 4*(2014), 18–21.

Roggema, R., Pugliese, A., Drissen, M., & Broekhuis, B. (2014) The Foodroof: How Cantagalo and Pavão-Pavãozinho Favelas Grow Their Own Food. In: R. Roggema & G. Keeffe (Eds.). (2014) *Why we need small cows. Ways to design for urban agriculture* (pp. 207–230). Velp: VHL Press.

Rotterdam, Gemeente. (2012). *Stimuleren Stadsland-bouw in en om Rotterdam*. Rotterdam: Gemeente Rotterdam.

Stadslandbouw, Expertisegroep. (2009). *Manifest Eetbaar Rotterdam*. Rotterdam: Expertisegroep Stadslandbouw.

Stadslandbouw, Stedennetwerk. (2010). *De Re Rustica Urbana Tandem Fit Surculus Arbor, of hoe een hype perspectief kan krijgen*. WUR: Wageningen.

Stutterheim, E. (2013). *De Eetbare Stad. Een onderzo-ek naar de mogelijke bijdrage van Stadslandbouw op de kwaliteit van de leefomgeving*. Groningen: RUG.

Van der Bie, R., Hermans, B., Pierik, C., Stroucken, L., & Wobma, E. (2012). *Smakelijk weten, Trends in voeding en gezondheid*. Den Haag/Heerlen: Cen-traal Bureau voor de Statistiek.

Van der Sande, B. (Ed.). (2012). *Food for the city. A future for the metropolis*. Rotterdam: NAi publish-ers/Stroom Den Haag.

Van Rossum, C. T. M., Fransen, H. P., Verkaik-Kloost-erman, J., Buurma-Rethans, E. J. M., & Ocké, M. C. (2011). Dutch National Food Consumption Survey 2007–2010. Diet of children and adults aged 7 to 69 years. *RIVM-Rapport 350050006*. Bilthoven: RIVM.

Van Rossum, C. T. M., & Geurts, M. (2013). Hoeveel mensen voldoen aan de Richtlijnen goede voed-ing? In: Volksgezondheid Toekomst Verkenning, Nationaal Kompas Volksgezondheid. Bilthoven: RIVM, http://www.nationaalkompas.nl. Nationaal Kompas Volksgezondheid \Determinanten\Leefstijl\ Voeding.

Veen, E., Breman, B., & Jansma, J. E. (2012). *Stadslandbouw. Een verkenning van groen en boer zijn in en om de stad*. Lelystad: Wageningen UR.

Viljoen, A. (Ed.). (2005). *CPULs. Continuous produc-tive urban landscapes*. Oxford, Burlington: Elsevier Architectural Press.

Voedingscentrum. (2014). Vlees. Retrieved July 24, 2014 from: http://www.voedingscentrum.nl/encyclo-pedie/vlees.aspx

Xie, E. (undated) *Tokyo train station opens rooftop garden to community*. http://garakami.com/20140411/tokyo-train-station-opens-rooftop-garden-to-community/

Chapter 3
Networks and Chains in the Provision of Food

Rob Roggema and Jeffrey Spangenberg

3.1 Introduction

Primarily due to economies of scale and globalization food chains have become extremely long, complex and opaque. In recent years there is trend in which consumers, retailers, craftsmen and producers are making a serious effort to change the status quo for number of different reasons, which are either business driven (price, efficiency, marketing, strategy), or less economy driven (sustainability, local economy, customer experience, circular economy). With the disconnection between production, distribution, preparation and consumption of food the market has alienated itself form all its participants. This feeling of estrangement and inconvenience seems to be substituted by several emerging trends, especially significant and visible in an urban context. Indeed, the fortress walls of the spatial separation of functions are showing some cracks. The changes are manifested in an explosive increase in mobile semi-permanent places where food is prepared and consumed at the same time, while on the other hand more attention is given to the production of food in the city. These flexible new concepts in the food services, illuminated through the rise of street food business, have changed the food landscape dramatically Not only in terms of need and demand, delivering more choice, ready to go concepts and a new food experience, but they also seem to represent a new generation of food

This chapter, in slightly different form, has previously been published in the proceedings of the 51st ISOCARP conference, 19–23 October 2015, Rotterdam, the Netherlands.

R. Roggema (✉)
Faculty of Design, Architecture and Building, University of Technology Sydney, Ultimo, NSW, Australia
e-mail: rob@cittaideale.eu

J. Spangenberg
Spang31, Urban Green and Urban Food, Food Night, Smaakboulevard, Amsterdam, The Netherlands

R. Roggema (ed.), *Food Roofs of Rio de Janeiro*,
DOI 10.1007/978-3-319-56739-6_3

entrepreneurs who really like to make in change in the social, cultural en environmental aspects of food. Accordingly they put a stronger emphasis on fresh, local and divers food. In that sense they coincide with emerging trends in urban agriculture.

These two typical urban developments, street food vendors and urban food growing, seem desperately seeking new connections. Some do even claim to have found new solid ground to shortcut the existing food chain. This existing chain 'from resource to mouth' is in this chapter compared with a new model of integrated networked regional planning of the food chains. This new model actually follows the new roadmaps set out by trendy chefs, local entrepreneurs, concerned citizens or local alderman.

3.2 Problem

Theoretical concepts, such as urban metabolism, and the practice of urban food production seem to be wide apart. One reason for this is a difference in the scale they seem to be working on and the lack of spatial planning as integrating platform at a scale that is applicable for both practical projects and sustainable theories. If we could spatially link resources, projects, waste flows and consumers at the level at which the reuse of flows is best possible, the efficiency of urban metabolism will be enhanced. Therefore it is necessary to define the interlinking scale, which is not too large to lose interest of individuals and local groups, but is also big enough to close cycles in a profitable way. In this chapter the City-Region scale is proposed as the spatial platform for sourcing resources, produce food, to process and transport, sell and consume it, and reuse streams of waste/materials, nutrients, water and energy.

First the chain of production, preparation and consumption is analysed. Secondly, the type of actors and spatial conditions are formulated, followed by an analysis of the flows in the Amsterdam West region. New thinking in networks is then applied to the case study and the chapter ends with conclusions.

3.3 The 'Production-Preparation-Consumption' Chain

Currently, many urban farming projects are (by definition) small-scale enterprises, often driven by non-profit social entrepreneurs and/or social collectives. In that sense these initiatives are not economically driven, hence they lack the stimuli to exchange resources and products to make profit. People primarily indulge in developing their (collective) gardens, but exchange of the products with others is merely 'collataral', with some exceptions (in the Netherlands: Marqt, Landmarkt, La Place and DeliXL). Some knowledge is exchanged, for instance in the Stedennetwerk Stadslandbouw (http://www.stedennetwerkstadslandbouw.nl), or the Urban Agriculture Linkedin Group (https://www.linkedin.com/topic/urban-agriculture), but the exchange of resources and products is very limited.

On the other side, the theory on local/circular economy, urban metabolism and cradle-to-cradle approaches is rapidly increasing. This marks a divide in a world of cherishable local practices and a world of ambitious sustainability concepts. The question is if these two worlds are really separate. And if so, how they can be linked. The circular economy (WEF 2014; Preston 2012; Wijkman and Skånberg 2015; Ellen MacArthur Foundation 2013), cradle-to-cradle (Braungart and McDonough 2002; Braungart et al. 2007), and urban metabolism (Ferrão and Fernández 2013; Kennedy et al. 2011; Barles 2010; Niza et al. 2009; Gandy 2004) are all theories that aim to use resources more efficient, recycle and re-use waste flows, and close cycles. Urban metabolism is seen as the future framework for strategic decision-making in spatial planning and the society (Hajer and Dassen 2014). This is not only important from a sustainability point of producing less waste, it is also an important economical factor as waste flows and used materials are valuable as new resources. However, the majority of literature is still theoretical, with minor links to practical applications.

The concrete, bottom-up and lower scale trends of Urban Agriculture is mostly disconnected from the abstract, theoretical principles of urban metabolism. It seems to be undisputed the urban environment is the only link between Urban Agriculture and Urban Metabolism. Another more compelling reason to look at the urban food chain lies in the fact that over 50% of the global population currently lives in cities (UNFPA 2007), which is expected to increase to 66% in 2050 (UN 2014), or even higher in some areas. The question how the urban population arranges its food is a complicated question. However, agricultural structurists claim to be able to feed the global population, even if it reaches nine billion people. The proposed agricultural system is a continuation of the current system, with large-scale production, efficiency, and distribution of food at world scale as the key characteristics. The question is whether this ideology is sustainable. It has numerous negative impacts, on food safety, food security, the environment, fossil fuel depletion, and more. The existing food flows around the world demonstrate painfully the absurdity of food travel, for example grain (Fig. 3.1).

Research at the Dutch national level shows that the total consumption in the country requires a surface of three times the Netherlands hence most of the consumption products come from other countries and continents (De Blois et al. 2014). The flipside of this is that most of the meat in the Netherlands is produced for export reasons and leaves the country. On the contrary, the fish the Dutch consume comes for the main part from other countries (De Blois et al. 2014; Hajer and Dassen 2014, Fig. 3.2). These statistics resonate with the conclusions drawn in the LEI-report how the Netherlands could feed its inhabitants without in- or export. In this scenario the diet has to change dramatically, with complete absence of meat (pork), the limited amount of grain used to feed the population and with a prominent role of chicken, eggs and potatoes in the diet (Terluin et al. 2013).

The analysis of the Dutch Food Chain (De Blois et al. 2014, Fig. 3.3) shows that 65,000 farmers produce for 6500 food manufacturers (processers), who deliver their products to 1500 suppliers. These suppliers deliver the products to only five purchasing companies, who deliver their goods to 25 supermarket concepts and 4400

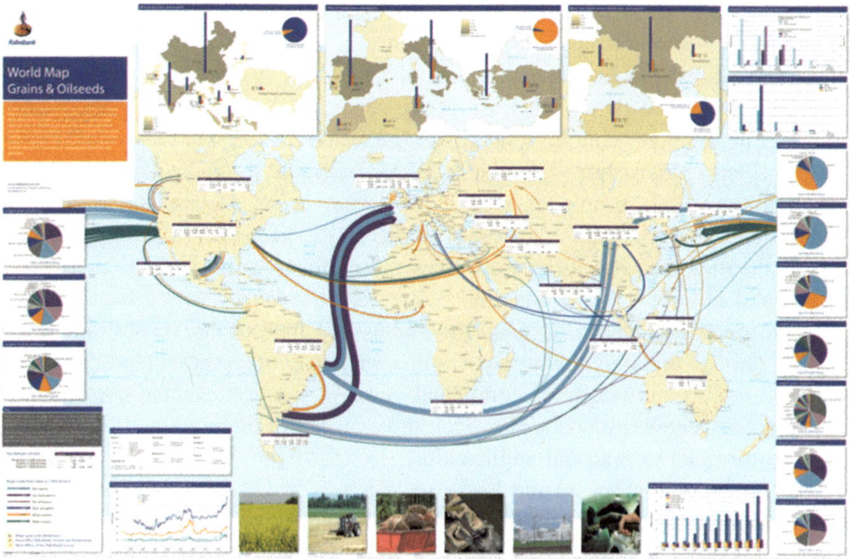

Fig. 3.1 Flows of grain around the world

supermarkets. The seven million shoppers buy their daily food there to feed the total of 16.5 million inhabitants. This chain shows an uneven power distribution, concentrating nearly all the power with five buying agencies, which decide about the distribution of the goods over the supermarkets. Secondly, a very thin line at the top of Fig. 3.3 shows the very few direct links between farmers and shoppers. A link that is truly underestimated and especially important for people without the budget to buy food from the supermarket. The FoodRoof is an important example to expand this relationship, represented by this thin line.

This illustrates the length of the chains and the disrupted power balances in them. Therefore, many scholars pledge for a model in which the food chains are shortened, and minimised to the urban or regional level (Aubry and Kebir 2013; The New York Academy of Sciences 2009; Santini and Gomez y Paloma 2013). The benefits of this alternative model for the food chains are multiple. Not only in terms of better insights in the quality and origin of food, less transportation miles and less environmental impacts, but also in social and cultural terms.

However, at which scale the food chains can be closed is complicated and differs for every product, crop, cultivation and market. Several food-planning scales are currently in use (Table 3.1).

1. The City Region food system encompasses the complex network of actors, processes and relationships to do with food production, processing, marketing, and consumption that exist in a given geographical region that includes a more or less concentrated urban centre and its surrounding peri-urban and rural

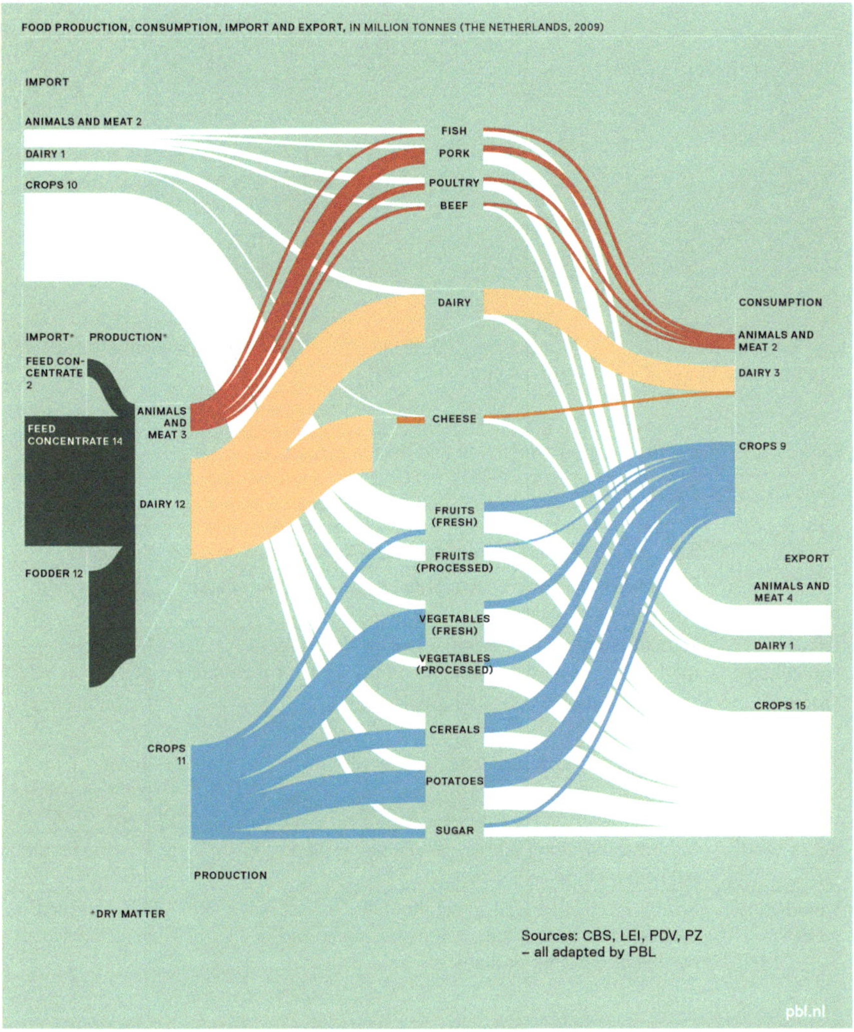

Fig. 3.2 Import, export, production and consumption in the current food system of the Netherlands (Hajer and Dassen 2014)

hinterland; a regional landscape across which flows of people, goods and ecosystem services are managed (FAO and RUAF 2015).

2. Food system planning is seen as an urban system (Pothukuchi and Kaufman 1999, 2000), but on the other hand the local scale is not the only scale to look at the food system (Born and Purcell 2006), as the system is scalable, and can be analysed at higher scales, even global.
3. Urban Agriculture is defined as: ‘an industry located within (intra-urban) or on the fringe (peri-urban) of a town, an urban center, a city or metropolis, which

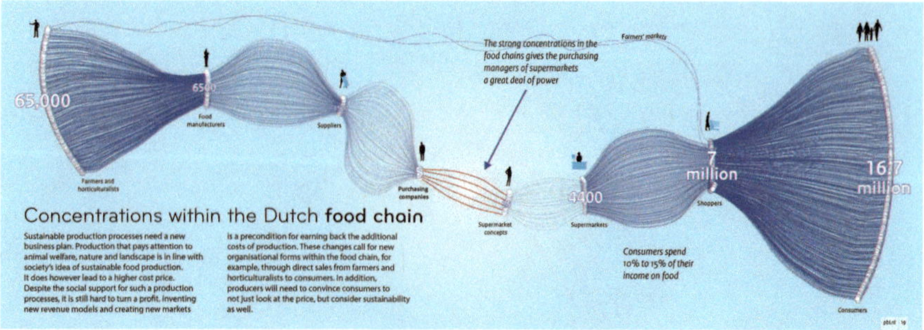

Fig. 3.3 The Dutch food chain in numbers (De Blois et al. 2014; http://www.pbl.nl/sites/default/files/cms/afbeeldingen/009i_ovg12_en.png)

Table 3.1 Types of urban food production and their typical scale

Type	Definition	Scale
City region	A more or less concentrated urban centre and its surrounding peri-urban and rural hinterland	Regional landscape
Food system planning	Planning of the food system at the urban or the local scale. The system is scalable	Urban-region
Urban agriculture	Agriculture within (intra-urban) or on the fringe (peri-urban) of a town, an urban centre, a city or metropolis	Urban and peri-urban
Urban farming	Farming within and around urban centres	Urban centres
Street food	Food sold in a street or other public place	Street, public/private space
Street vendor	Person selling food on the pavements or other public/private areas, or mobile	Pavement, public area
Community garden	Shared productive land in neighbourhoods, schools, connected to institutions such as hospitals, and on residential housing grounds	Piece of land in neighbourhood

grows or raises, processes and distributes a diversity of food and non-food products, reusing mainly human and material resources, products and services found in and around that urban area, and in turn supplying human and material resources, products and services largely to that urban area' (Mougeot 1999).

4. Urban farming is the growing, processing, and distribution of food or livestock within and around urban centers with the goal of generating income (Thoreau 2010; Poulsen and Spiker 2014).
5. Street food is ready-to-eat food or drink sold in a street or other public place, such as a market or fair, by a hawker or vendor, often from a portable food booth, food cart or food truck (Simopoulos and Bhat 2000).

6. A street vendor is a person who offers goods or services for sale to the public without having a permanently built structure but with a temporary, static structure or mobile stall (or head-load). Street vendors could be stationary and occupy space on the pavements or other public/private areas, or could be mobile, and move from place to place carrying their wares on push carts or in cycles or baskets on their heads, or could sell their wares in moving buses (MHUPA 2004; Sundaram 2008).
7. Community garden/consumer collectives: A community garden is any piece of land gardened by a group of people, utilizing either individual or shared plots on private or public land. The land may produce fruit, vegetables, and/or ornamentals. Community gardens may be found in neighbourhoods, schools, connected to institutions such as hospitals, and on residential housing grounds (Bauermeister et al. 2010).

Network enterprises, such as Air-bnb (www.airbnb.com), Uber (www.uber.com), With Local (www.withlocals.com), Thuisafgehaald (www.thuisafgehaald.nl), and connectivity will in the near future become more important than locality. This implies that networks and the quality (value) of products will determine the price more than distance and the cost of resources. Therefore, it will become beneficial to deliver high value over short distances. For the planning of the food chain, networks can form the linking pin between the conceptual theories of metabolism and the practice of local food producers, vendors and consumers.

3.4 The Actors

Within the City-Region (see Chap. 2) many visible and invisible actors are active and all of them are producing, using or processing resources and products, and waste. At a systemic level the actors are all connected. Currently many of these actors recruit their resources and products from outside the city region, local producers are isolated within and metabolism concepts theoretically use the urban scale as the platform for their deliberations. In order to connect these issues firstly the flows and relationships must be identified, including the resourcing, production, transportation, processing and consumption of flows related to food. The system arising from this is shown in Fig. 3.4.

In this figure the resources of clean water, renewable energy, seeds and nutrients are, ideally from within the city region, provided to the producers. They need these resources and deliver products, such as vegetables, herbs, fruit, fish, meat. The producers deliver their products, which are distributed and processed by the producers themselves or distribution-entrepreneurs. Consumers, specialised processing companies or the producers themselves take care of processing the produce in the form of processed food or meals. Finally, these products are sold to consumers or vendors, such as street vendors, markets or restaurants.

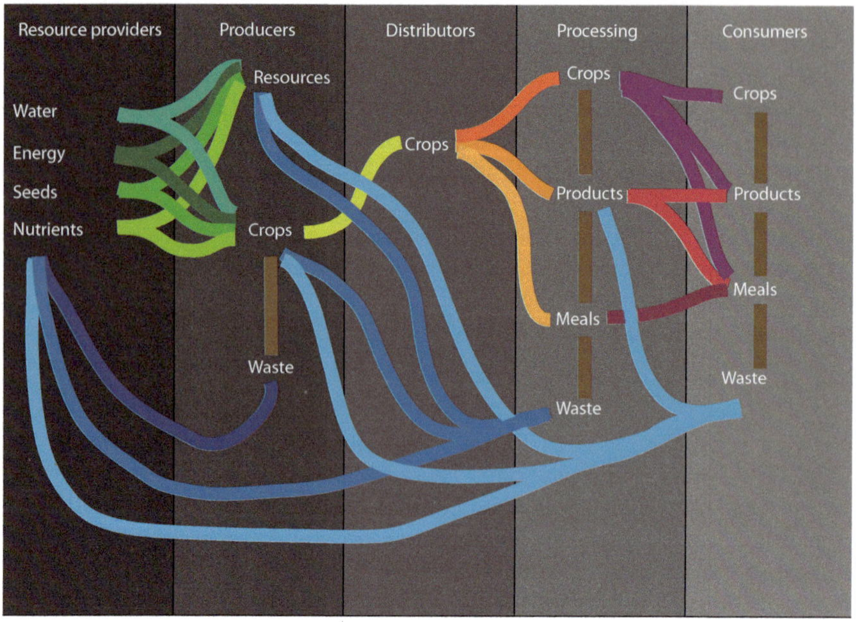

Fig. 3.4 Actors and their exemplary connectivity of flows

In this diagram the following actors appear:

1. The resource provider is a person or institution that provides the resources to the producers. The resources are water, energy, nutrients and seeds or seedlings.
2. The producer is the person or enterprise, such as a farmer or land-owner, that produces crops, such as vegetables, fruits, herbs, but also, cattle or fish.
3. The distributor is the organization that takes care of the transportation of the crops to processing enterprise or directly to markets or consumers. The producer and distributor can be the same entity.
4. The processor is the person or enterprise that uses the crops to add value and sells them. Processing of crops into products, such as jam, sausages, bread and many others takes place in process industries, or at home, while the processing of crops into meals occurs in restaurants or in street food booths.
5. The consumer is the person or group of persons that is consuming the crops, meals or products. Consumption takes place at home, in restaurants or at street food booths. Consumers, processors and producers each have waste flows, which ideally are recycled back into the system.

The next step in concretising this abstract connectivity scheme can be conducted when the actors include their demands and deliveries are scrutinized.

3.5 Spatial Conditions

When the Tuinen van West projects (www.tuinenvanwest.info, Innovatie-programma Stadslandbouw Tuinen van West 2010), the Meervaart urban design (Roggema 2015) and De Hallen food court (Building Business 2014; FoodHallen Amsterdam, n.d.), which are all located in Amsterdam, are analysed, some typical, but diverse, spatial conditions can be formulated (Table 3.2).

Once the existing actors are known and located, the spatial conditions of these actors (Table 3.2) give insights in the potential suitable locations in the city. On a map the current and future producers, consumers and street vendors, can be located once the spatial requirements are combined with the spatial capacity in the city in order to locate and identify the best possible fits between actors and urban context (see Chap. 2). When this map is subsequently extended with the size of the flows between the actors, the whole model of food production with closing cycles on a (sub-) regional city scale can be constructed.

3.6 Analysis of Current Flows

When the conceptual diagram of food flows (Fig. 3.4) is translated in a concretized systemic analysis of existing, classic flows at regional scale (Fig. 3.5), it is clear that most of the cycles are not circular. The scheme shows the gaps in the circle: the resources are coming from outside the system (by plane or ship via airports or harbours, and from regional suppliers such as power plants and water companies. On the other hand the waste flows, coming from producers, restaurants, markets and

Table 3.2 Approximate sizes and spatial requirements food actors

Actor	Size	Requirements
Producers	50 × 100 m	Place to recycle water, energy, nutrients, waste
Markets	30 × 70 m	Central in neighborhood
Restaurants	10 × 20 m	Location
Street vendors	2 × 5 m	Food truck, booth, garage
Consumer collectives	20 × 30 m	Central in community, collective shed, variety of crops, space to recycle water, energy, nutrients, waste
Flows		
Water	2 × 2 m, 4 × 5 m	Water tank, pond
Seeds/seedlings	10 × 5 m	Storage, shed/nursery
Energy	10 × 20 m	Biogas, windmill or solar panels
Nutrients	2 × 3 m	Composting
Food-distribution	2.5 × ∞, 7.5 × ∞	Bike lane, street

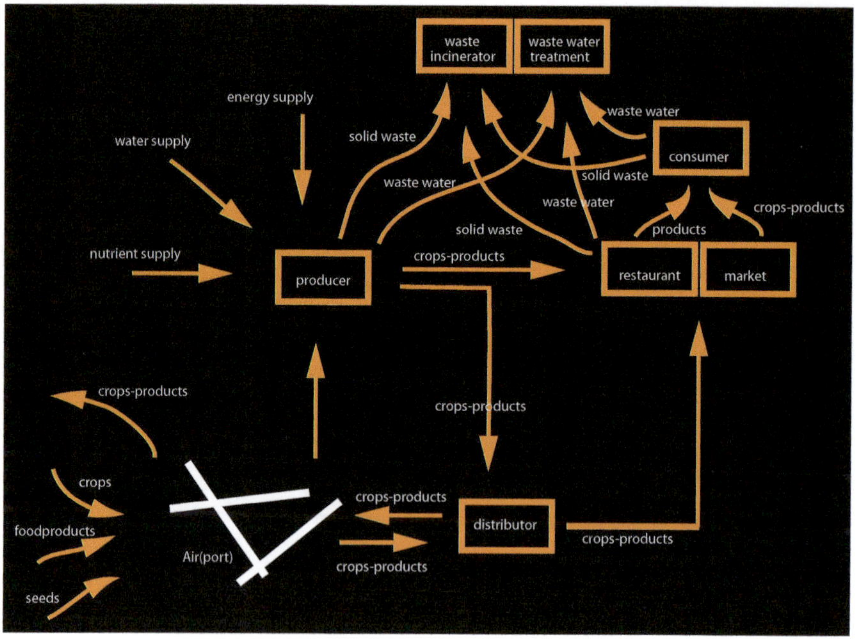

Fig. 3.5 Current network of flows

consumers, are treated centrally at regional scale in water treatment plants and waste incinerators, and are not recycled and re-used as resources.

- Crops, food-products and seeds are imported in the harbour or the airport, and directly delivered to producers or though the intermediary of a distributor to restaurants and markets.
- The producers of food crops receive their nutrients, water and energy from regional companies, such as Waternet or PWN Waterleidingbedrijf Noord-Holland (water), Nuon, E-On, Qurrent, Greenchoice, Oxxio and others (energy), and Mesthandel CG Kroon (nutrients).
- The producers deliver crops and products to restaurants and markets (Albert Cuyp, Dappermarkt, Ten Cate markt) and to the distributors (e.g. Vroegop-Windig, Dawn Amsterdam) who deliver these to restaurants and markets, but also export the products through the ports.
- The producers discharge their waste to the water treatment plant (RWZI Amsterdam West) and the waste incinerator (Afval Energie Bedrijf Amsterdam).
- The consumer obtains its products from restaurants or markets, or directly from the producer and produces waste flows that are treated in the water treatment and waste incinerator.

In practice, the current system leaks. It leaks carbon in the generation of energy, using fossil resources, it leaks carbon during transportation bringing the products to the Amsterdam region and in distributing these over the region, and it leaks wastewater and solid waste during processing and preparation of food. A substantial amount of resources, crops and products originates from all over the world.

3.7 Future Thinking About Networks

Historically our discourse of food chains in dominated by linear thinking. Goods or resources went into a system, were processed and transformed in something else, and subsequently left the system without feedback loop. This is a centralised type of system. Thinking about these chains is changing and they are increasingly seen as a network of functions, relationships, producers and consumers, in which every actor could change roles. Functionalities of these systems are distributed over the network, reason why these types of network is called a distributed network (Biggs et al. 2008a, 2010; Roggema and Stremke 2012). Distributed systems are characterised by multiple and mutual interactions and nodes, and the closing of cycles at a local scale (Fig. 3.6).

Research has applied distributed network thinking to distributed water systems (Biggs et al. 2008b; Mays 2000; The Johnson Foundation at Wingspread 2014), and

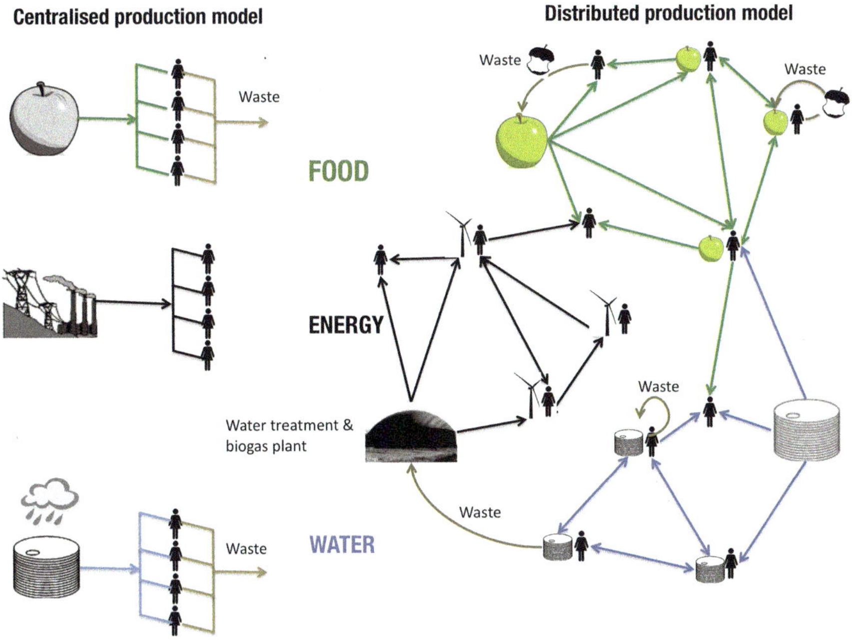

Fig. 3.6 Centralised versus distributed systems (VEIL, http://www.ecoinnovationlab.com/website/wp-content/attachments/dist-sys-model.png)

distributed energy systems (McCormick et al. 2008; Alanne and Saari 2006; Esmaili et al. 2005), but so far it only touches marginally on the food topic.

When the principle of distributed network is applied to food production at the city region scale, the main actors are the producers, consumers, markets, restaurants and street vendors (Fig. 3.7). Producers and consumers provide in their own resources of water, energy, nutrients, seeds, and recycle their waste. Each of the actors is connected with every other actor. The connection represents the flows to and from each actor.

In this model (Fig. 3.7):

- P = Producer, arranging its own S (Seeds), Wa (Water), E (Energy), N (Nutrients), Ws (Waste treatment)
- C = Consumer or Consumer collective, arranging its own S (Seeds), Wa (Water), E (Energy), N (Nutrients), Ws (Waste treatment)
- The sellers are: MKT = Market
- RST = Restaurant
- SV = Street Vendor.

In Fig. 3.7 the producing capacity of Consumer collectives is left out of the scheme because of readability. The next step is to identify the actors on the map and relate them to each other. The actors, subdivided in producers, collective vegetable

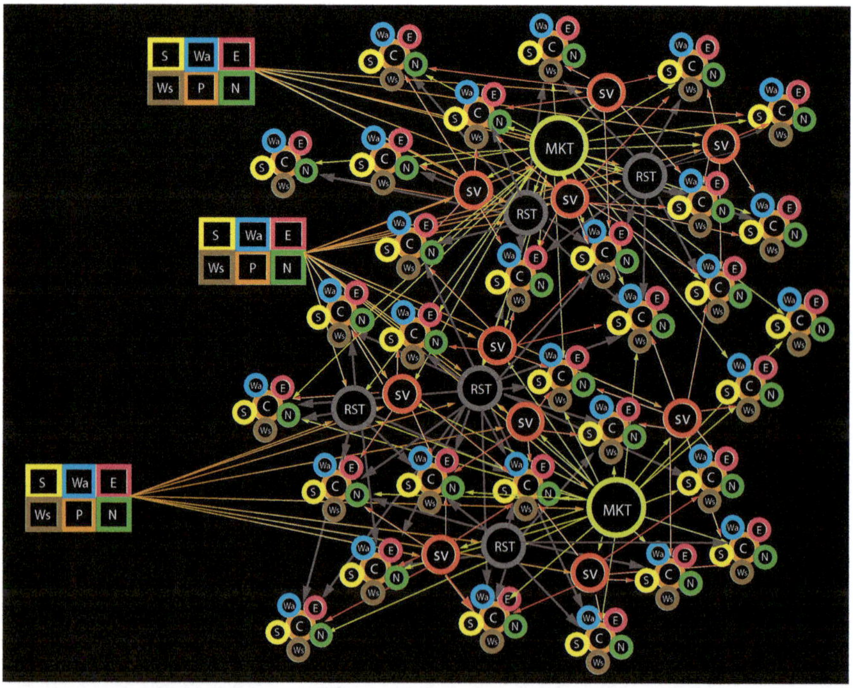

Fig. 3.7 Potential future network of flows between producers, consumers and sellers

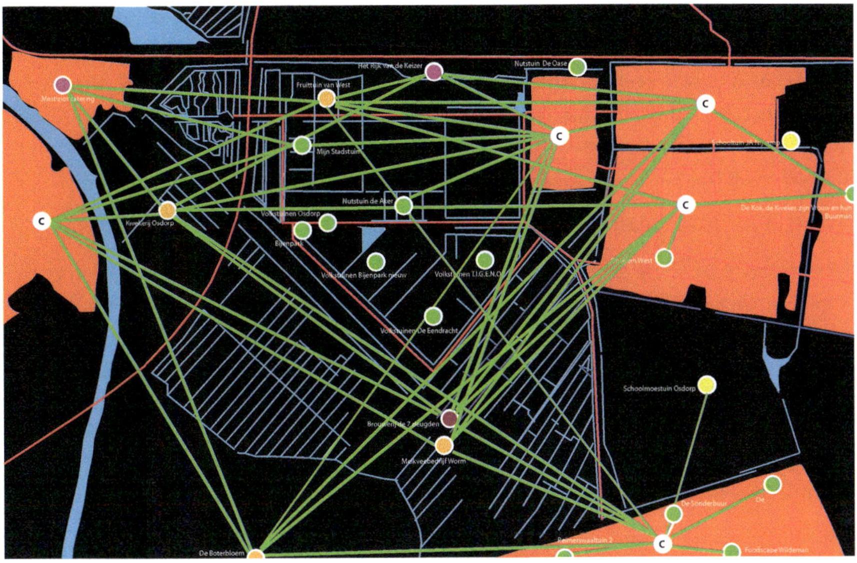

Fig. 3.8 Connections between producers, consumers and processors in the Tuinen van West, Amsterdam

gardens, school gardens, restaurants and consumers, are shown on the map of Tuinen van West, Amsterdam (Fig. 3.8), including their relationships (supply, provision, purchasing).

- The producers, such as Fruittuin van West and De Boterbloem, use their own generated water, energy, seeds and nutrients, and their recycled waste, in their own production process and deliver products and crops to markets, restaurants and street vendors, or even directly to consumers.
- The markets, such as Boerenmarkt Haarlemmerplein and Mercatormarkt receive their products from producers and deliver products and crops to restaurants, street vendors and consumers.
- Street vendors (e.g. Boer Geert, Vleesch noch Visch, Tho Vietnamese Loempia's) and restaurants (e.g. Het Rijk van de Keizer, Ivo's Kitchen) receive their crops and products from producers and, some, from consumer collectives (e.g. Reimerswaalbuurt 1 and 2, Wijsgerenbuurt, De Dovenetel, De Kok, de Kweker, zijn Vrouw en hun Buurman, Hanno Klein tuin, De Groene Vaart; Lems and Van der Veen 2011) and deliver their products as meals to consumers.
- Consumers or consumer collectives use their own generated energy, nutrients, water and seeds, and their recycled waste to grow crops, which they consume themselves and deliver to restaurants and street vendors.
- The restaurants, street vendors and markets produce waste, which is delivered back to the producers in the form of nutrients and fertilizer.

This alternative model for a food chain includes distributed systems for energy, waste, nutrients and water and closes the cycles at the lowest possible scale. The system as a whole could well function at the scale of (a part of) the city region, such as in this case study, the western fringe of Amsterdam. The next step in the research will be to quantify the flows in this model and see whether this model can lead to a closed system.

What does this mean for the FoodRoofs and the food system in favelas?

The Food Chain in favelas could, despite being very different, follow the same systemic reasoning. For a favela, such as Cantagalo the food chain ends before it can even enter the area. Therefore it is essential to connect the favela to the food chain through a system of FoodRoofs. At the level of the favela, a connected system of roofs grow food, and form at the same time a distributed water and energy system (see Chap. 6). Together they are a relatively large entity, which is interesting for producers, processors and consumers within the food chain, but located outside the favela.

3.8 Conclusions

A different way of organising the food chain at the urban level has lead to multiple insights:

Cities tend to continue their business as usual practices, and seem to underestimate the potential benefits of a distributed food system, equal to the recent developments in distributed energy generation in so-called smart-grids.

The role of consumers as participants in the food producing process, for instance as entrepreneurs or food growers is undiscovered yet, but a promising opportunity to grow more food in urban environments and to involve residents in the production process, making them aware of the existing challenges and opportunities.

Cities should start their spatial thinking about food production from the local level upwards instead of assuming the global food system is the standard along which lines they organize their food supply.

The street food vendors and even so the owners of a FoodRoof, play a crucial role in bringing together the local production of food (which they need for their business), the preparation of food (what they do for a business), and the consumption of food (which they provide by their business).

Street food vendors also mark a change in how the community relates to food, as they experience the production and preparation more directly. The street food vendors are the frontrunners in making this experience manifest.

The scale at which a distributed system of food production operates is not yet clear. However, the sub-urban scale seems the place where many elements of this system come together. The local production, the local market, the entrepreneurs, the providers of resources, the freshness of products and the easy local transportation all play a crucial role and are all located conveniently at the scale of the urban sub-centre at the fringe of large cities and metropolises.

The planning profession should embrace thinking in distributed food systems, as it is by definition an approach in which local and regional planners could contribute.

- The localisation of producers or resources, distributors, the growers of food, the entrepreneurs and the consumers requires spatial planning, not only in a logistic manner but also in the perspective of giving urban transition zones the spatial quality they deserve.
- Beyond direct land-use planning, planners could also investigate the sizes and volumes of the diverse flows in these systems and identify the most optimal locations for different components in relation with each other.
- Planners are concerned with the planning of infrastructure accommodating the flows to flow.
- Finally, the planners' profession is necessary allowing and stimulating these food growing activities and initiatives at urban, sub-urban and urban fringe levels for instance by regulation, in which incentives are given to localize the production and use of resources for the purpose of growing food.

Each neighbourhood or urban precinct could apply the principles elaborated in this chapter. Given the local specific circumstances the distributed system and the flows of resources, produce and waste will be different. However, for every precinct the application of a model as proposed in this chapter will minimise food-miles, provide healthier and fresh food and give local entrepreneurs the chance for economic sustainability.

The benefits of such a new model of distributed and integrated network for food production and distribution are:

- A better tuned exchange of flows
- Less transportation, especially at global level
- Production and consumption within one sub-region
- Easier control and securing of food
- More engagement of local people.

In practice it may still be difficult to shift to this new model, as food production is part of the global market, and price competition is fierce, but on the other hand people in cities no longer take their food for granted and want to know who, where, what and with which resources their food is produced. The distributed network model offers this information by nature.

Current research on this new model is still scarce and a more research is necessary to gain deeper insights in the benefits of this model. Therefore specific research should investigate:

1. The optimal size at which the distributed network model operates best
2. The size and connections of the flows of water, energy, seeds and nutrients, and waste, including their spatial conditions

3. The differences with the existing food chains
4. The optimal process of (spatial) planning and distribution of food producers, processers and consumers.

However, the first next step is to make an inventory of the concrete actors, including the sizes of their respective resources and products.

References

Alanne, K., & Saari, A. (2006). Distributed energy generation and sustainable development. *Renewable and Sustainable Energy Reviews, 10*(6), 539–558.

Aubry, C., & Kebir, L. (2013). Shortening food supply chains: A means for maintaining agriculture close to urban areas? The case of the French metropolitan area of Paris. *Food Policy, 41,* 85–93.

Barles, S. (2010). Society, energy and materials: The contribution of urban metabolism studies to sustainable urban development issues. *Journal of Environmental Planning and Management, 53*(4), 439–455 (Special Issue: New Perspectives on Sustainable Development). doi:10.1080/09640561003703772

Bauermeister, M., Swain, S., & Rilla, E. (2010). *Marin county community garden needs assessment.* Oakland: University of California Cooperative Extension-Marin. http://cemarin.ucdavis.edu/files/66838.pdf

Biggs, C., Ryan, C., & Wiseman, J. (2008a). *Distributed systems: A design model for sustainable and resilient infrastructure.* Melbourne: Victorian Eco-Innovation Lab, University of Melbourne.

Biggs, C., Ryan, C., & Wiseman, J. (2010). *Localised solutions: Building capacity and resilience with distributed production systems.* Melbourne: Victorian Eco-Innovation Lab, University of Melbourne.

Biggs, C., Ryan, C., Wiseman, J., & Larsen, K. (2008b). *Distributed water systems: A networked and localised approach for sustainable water services.* Melbourne: Victorian Eco-Innovation Lab, University of Melbourne.

Born, B., & Purcell, M. (2006). Avoiding the local trap. Scale and food systems in planning research. *Journal of Planning Education and Research, 26,* 195–207. doi:10.1177/0739456X06291389.

Braungart, M., & McDonough, W. (2002). *Cradle to cradle: Remaking the way we make things.* New York: North Point Press.

Braungart, M., McDonough, W., & Bollinger, A. (2007) Cradle-to-cradle design: Creating healthy emissions—a strategy for eco-effective product and system design. *Journal of Cleaner Production, 15*(13–14), 1337–1348.

Building Business. (2014). WKO in tramremise 'De Hallen'. Published online May 26, 2014. http://www.buildingbusiness.nl/news/1131/15/WKO-in-tramremise-De-Hallen/

De Blois, F., Brandes, L., Van Hoorn, A., Huitzing, H., Mercx, M., Nabielek, K., et al. (Eds.). (2014). *The Netherlands in 21 infographics. Facts and figures on the human environment.* Den Haag: PBL.

Ellen MacArthur Foundation. (2013). *Towards the circular economy. Opportunities for the consumer goods sector.* Ellen MacArthur Foundation: Isle of Wight.

Esmaili, R., Xu, L., & Nichols, D. (2005). An optimal design of a grid connected hybrid wind/photovoltaic/fuel cell system for distributed energy production. In *31st Annual Conference of IEEE on Industrial Electronics Society, 2005, IECON 2005, 6–10 November 2005.* IEEE. doi:10.1109/IECON.2005.1569298

FAO and RUAF. (2015). *City region food systems. Building sustainable and resilient city regions*. Rome: FAO, RUAF.
Ferrão, P., & Fernández, J. E. (2013). *Sustainable urban metabolism*. Cambridge, London: The MIT Press.
FoodHallen Amsterdam. (n.d.). Published online. http://www.foodhallen.nl
Gandy, M. (2004). Rethinking urban metabolism: Water, space and the modern city. *City: analysis of urban trends, culture, theory, policy, action, 8*(3), 363–379. doi:10.1080/136048104 2000313509
Hajer, M., & Dassen, T. (2014). *Smart about cities. Visualising the challenge for 21st century urbanism*. Rotterdam: nai010 publishers/PBL publishers.
Innovatieprogramma Stadslandbouw Tuinen van West. (2010). *Rapportage Fase 2 Ontwerp "Tuinen van West als basisschool"*. Amsterdam: Gemeente Amsterdam.
Kennedy, C., Pincetl, S., & Bunje, P. (2011). The study of urban metabolism and its applications to urban planning and design. *Environmental Pollution, 159*(8–9), 1965–1973.
Lems, E., & Van der Veen, W. (2011). *Inspiratierapport Buurtmoestuinen. Kansen voor sociale cohesie en leefbaarheid in Amsterdam Nieuw-West*. Amsterdam: Gemeente Amsterdam, stadsdeel Nieuw-West.
Mays, L. W. (2000). *Water distribution system handbook*. New York, San Francisco, Washington, D.C., Auckland, Bogotá, Caracas, Lisbon, London, Madrid, Mexico City, Milan, Montreal, New Delhi, San Juan, Singapore, Sydney, Tokyo, Toronto: McGraw-Hill; The McGraw-Hill Companies, Inc.
McCormick, K., Falk, R., & Viswanathan, S. (2008). *An overview of development in distributed energy with country, state and city examples from EU and USA*. Melbourne: Victorian Eco-Innovation Lab, University of Melbourne.
MHUPA. (2004). *National policy on urban street vendors*. Department of Urban Employment and Poverty Alleviation, Ministry of Urban Development and Poverty Alleviation. New Delhi: Government of India.
Mougeot, J. A. (1999). Urban agriculture: Definition, presence, potential and risks, main policy challenges. CFP report Series, Canada.
Niza, S., Rosado, L., & Ferrão, P. (2009). Urban metabolism. Methodological advances in urban material flow accounting based on the Lisbon case study. *Journal of Industrial Ecology, 13*(3), 384–405. doi:10.1111/j.1530-9290.2009.00130.x
Pothukuchi, K., & Kaufman, J. L. (1999). Placing the food system on the urban agenda: The role of municipal institutions in food systems planning. *Agriculture and Human Values, 16*, 213–224.
Pothukuchi, K., & Kaufman, J. L. (2000). The food system: A stranger to the planning field. *Journal of the American Planning Association, 66*(2), 113–124.
Poulsen, M. N., & Spiker, M. L. (2014). *Integrating urban farms into the social landscape of cities. Recommendations for strengthening the relationship between urban farms and local communities*. Johns Hopkins Bloomberg School of Public Health.
Preston, F. (2012). *A global redesign? Shaping the circular economy*. Energy, Environment and Resource Governance, Briefing Paper. EERG BP 2012/02. London: Chatham House.
Roggema, R. (2015). The reinvention of the academic conference: How active delegates develop productive city concepts. *Future of Food: Journal on Food, Agriculture and Society, 3*(1), 63–78.
Roggema, R., & Stremke, S. (2012). Networks as the driving force for climate design. In R. Roggema (Ed.), *Swarming landscapes: The art of designing for climate adaptation* (pp. 91–116). Dordrecht, Heidelberg, London: Springer.
Santini, F., & Gomez y Paloma, S. (2013). *Short food supply chains and local food systems in the EU. A state of play of their socio-economic characteristics*. JRC Scientific and Policy Reports, European Commission. Luxembourg: Publications Office of the European Union.
Simopoulos, A. P., & Bhat, R. V. (2000). *Street foods*. Basel: Karger AG.
Stedennetwerk Stadslandbouw. Linkedin Network Group: https://www.linkedin.com/grp/home?gid=3212564&trk=my_groups-tile-grp

Sundaram, S. S. (2008). National policy for urban street vendors and its impact. *Economic and Political Weekly, 43*(43), 22–25.

Terluin, I. J., Kamphuis, B. M., Oudendag, D. A., & Van Leeuwen, M. G. A. (2013). *Food supply in the Netherlands under exceptional crisis conditions (Summary)*. The Hague: LEI Wageningen UR. Report 2013-012.

The Johnson Foundation at Wingspread. (2014). *Optimizing the structure and scale of urban water infrastructure: Integrating distributed systems*. Racine, WI: The Johnson Foundation at Wingspread.

The New York Academy of Sciences. (2009). Shortening the food chain. Farming in the city. eBriefing. http://www.nyas.org/Publications/Ebriefings/Detail.aspx?cid=00bbf00c-d2e5-4681-9b5d-6a3732ac303c

Thoreau, C. M. (2010). Defining urban farming. Vancouver Urban Micro. Published online. http://urbanmicro.ca/2010/10/15/defining-urban-farming/

UN. (2014). *World urbanization prospects*. New York: UN.

UNFPA. (2007). *State of world population 2007. Unleashing the potential of urban growth*. New York: UNFPA.

Urban Agriculture. Linkedin Network Group: https://www.linkedin.com/grp/home?gid=1872242&sort=recent&trk=my_groups-tile-flipgrp

Urban Farming. Linkedin Network Group: https://www.linkedin.com/grp/home?gid=3428863&trk=my_groups-tile-grp

WEF (World Economic Forum). (2014). *Towards the circular economy: Accelerating the scale-up across global supply chains*. Geneva: WEF.

Wijkman, A., & Skånberg, C. (2015). *The circular economy and benefits for society Swedish case study shows jobs and climate as clear winners*. Winterthur: The Club of Rome.

Author Biography

Jeffrey Spangenberg Spang31, Entrepreneur in Urban green and urban food; Owner; Food Night©, Smaakboulevard©

Part II

Chapter 4
Policy Context: PAC

Ruth Jurberg

4.1 Introduction

Brazil is predominantly considered an urban country, according to the Brazilian Institute of Geography and Statistics (IBGE), the urbanisation rate increased from 44.7% in 1960 to 84.3% in 2010 (IBGE 2010) (Fig. 4.1).

The 1940s and 1950s were marked by migratory movements originating mainly in the North Eastern states of Brazil and moving to the Southeast and South regions. Migration was encouraged through better access to consumer goods, employment and the illusion of a better life in the southern areas of Brazil. Without any form of official support and thickening the informal market, this newly arrived population faced serious problems, amongst which with their housing supply.

The situation of precariousness and the increase of the number of slums worsened from the 1960s, when Brazil went through an urban explosion, which brought a large number of rural workers, who had no alternative in the country side, to live in the peripheral areas and slums of the cities. In the same period the cities expanded, the size of the slums increased dramatically, without hardly any public investment or monitoring. In the case of Rio de Janeiro, the government spent decades omitting the existence of these regions that grew without any form of planning. For decades these slums were 'out of the maps' and not in the registers from the public sector. As such, they didn't exist.

In this context, Brazilian cities produced and reproduced social inequalities on a daily basis and the slums are the most visible reflection of this inequality. These spaces were created through a self-construction system in 'non aedificandis' (unbuildable) areas as at riversides and canals or on the top of the hills. These spaces were of no interest to the real estate market.

R. Jurberg (✉)
State Government of Rio de Janeiro, Rio de Janeiro, Brazil
e-mail: ruthjurberg@gmail.com

R. Roggema (ed.), *Food Roofs of Rio de Janeiro*,
DOI 10.1007/978-3-319-56739-6_4

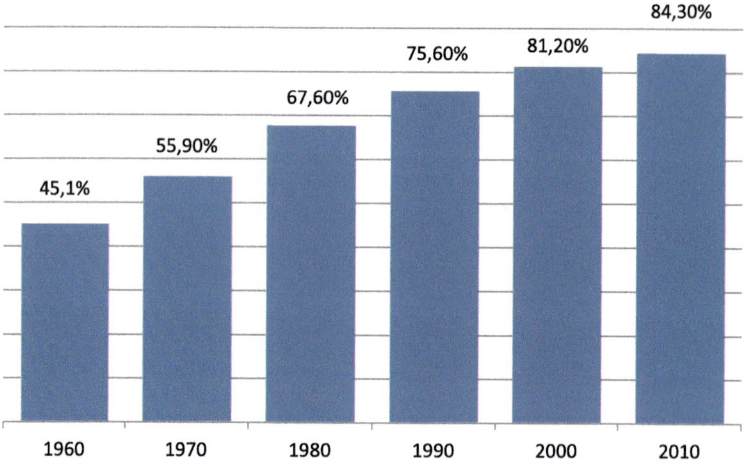

Fig. 4.1 The urbanisation rate (IBGE 2010; *Source* http://seriesestatisticas.ibge.gov.br/series.aspx?vcodigo=POP122)

These city neighbourhoods were built during an intense period of precarious living conditions, with as the main issues the high levels of underemployment and an informality in labour relations, low educational and health indicators, with economic and environmental indicators being below average, and a high population density and occupation of the drug traffic in these territories.

4.2 The PAC Program

In 2007 the Federal government launched the Growth Acceleration Program (PAC Program) for the entire country. The program promoted the resumption of planning and execution of major social, urban, logistics and energy infrastructure projects in the slums, contributing to its accelerated and sustainable development.

Designed as a strategic plan for rescuing planning and resumption of investments in structuring sectors, such as transportation system, infrastructure, housing units of the country, PAC has contributed decisively to increasing the supply of jobs and generating income, and has increased public and private investment in key projects (http://dados.gov.br/dataset/obras-do-pac-programa-de-aceleracao-do-crescimento).

In Rio de Janeiro State the Program included interventions in low-income communities such as Complexo do Alemão, Rocinha, Manguinhos and Pavão-Pavãozinho-Cantagalo and the Preventorio in Niteroi city (Fig. 4.2), totalling about 350,000 residents. The project is aimed at urbanizing communities, and is supported by large social works, whose main objective is to transform the physical interventions into sustainable development. The benefits of economic growth should

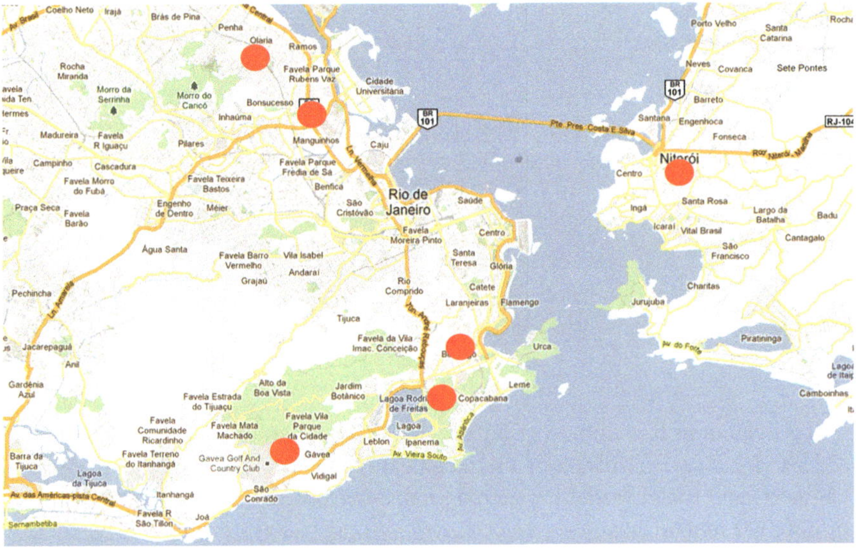

Fig. 4.2 Map of the interventions of PAC Program. *Source* PAC Program—map elaborated by EMOP

be distributed and support resources under democratic management. The initial investment was US $ 850,000,000.00 in development projects, construction of various community facilities such as high schools, kindergartens, health clinics, 24 h Hospital, Park Libraries, Youth Reference Centre and a cable car, a mass transportation system, which serves around 12,000 people/day in Complexo do Alemão, in addition to the construction of 3820 new housing units for low-income population. In just one year, more than 7500 families were part of reallocations processes in Manguinhos, Complexo do Alemão and Rocinha. In addition, the program included land regularisation actions for 5000 residences in Rocinha, 18,000 in the Complexo do Alemão and 4200 in Manguinhos.

The innovative program proposed to develop Urbanisation (Fig. 4.3), in which physical, social and legal components work jointly and interact throughout the project. The government improved its knowledge of the territory, optimized the division of resources and organised the commitment of the population.

Those low-income communities were chosen for implementation of the PAC Program in 2007, facing problems of urban and housing insecurity, unemployment, social vulnerability, violence, and drug trafficking, in combination with a low credibility of the actions of the government, which had been absent for many years. The housing in these areas was poor and there was a lack of infrastructure. There were no recreational areas nor access and the public transportation system was distant and precarious. There were no connections between the formal city (city districts) and the informal communities, creating more difficulties for hundreds of residents.

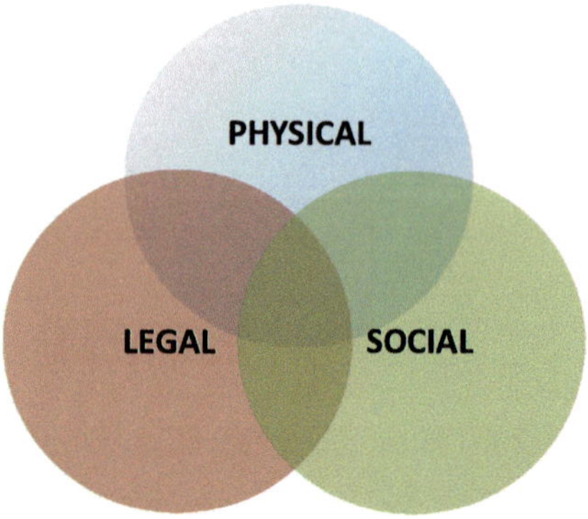

Fig. 4.3 Model for integrated urbanisation

To overcome this initial period in which the government was discredited and absent, a joint performance of the technical teams (architects, urban planner and engineer) and social workers were required. Selected communities had a population of 350,000 inhabitants.

The government chose to set up strong collaborations between social work urbanisation and land tenure, to ensure an on-going social dialogue and to improve, in an effective and permanent way, the quality of life for these people.

The work began with the definition of projects and prioritization of interventions, focusing on urbanisation of the communities, accessibility, infrastructure deployment, construction of public facilities and new housing.

One of the biggest challenges was to regain the credibility and trust of the population in the governmental program. For many years, the State Government was not able to enter the favelas with police forces, without ensuring the most basic rights of this population. The result was that the population was incredulous about the purpose of public investments in these areas, and is rewarded after this initial difficulty to redress the confidence, and implement projects involving democratic proposals of residents and community leaders.

It was decided by the Ministry of Cities that together with the progress of urbanisation the social work would be implemented. Initially the social work team carried out a diagnosis of the communities and a comprehensive house and business census in three of the slums: Complexo do Alemão, Manguinhos and Rocinha.

Almost 1500 residents from the three slums were hired to conduct this census that lasted one year and two months, between 2008 and 2009. This was an opportunity to understand what was going on in those areas and to deeply understand the social and economic situation of the population. It was an X-ray of these areas.

To carry out urbanisation works in the favelas, it was necessary to work jointly with the resettlement of thousands of families.

This resettlement included a total of 7500 families and the government offered three options to choose from: compensation, assisted or purchase new apartments. The proposal methodology created three committees that were formed: a relocation committee, monitoring work and income generation. The main idea is to guarantee the community participation on those committees that happened every two weeks. All the residents were invited to participate and during those meetings the technicians from the government (social workers and architects) and the residents debated issues concerned to the urbanisation of the slums, job opportunities during the reconstruction process and the problems concerned about the relocation process.

4.3 Characteristics of the Intervention Area at Pavao-Pavaozinho and Cantagalo

The Cantagalo Massif, is situated next to the neighbourhoods of Ipanema and Copacabana, in the southern zone of Rio de Janeiro (Fig. 4.4).

Although these areas are geographically very well located amidst a middle-class area, residents of these communities have always had similar problems to other areas with low development standards, such as flooding, open sewage, urban violence and housing.

The urban occupation that shaped Pavão, Pavãozinho and Cantagalo is one of the oldest in the City of Rio de Janeiro. The history of Cantagalo began before Pavão Pavãozinho early in the 20th century when the first inhabitants built their

Fig. 4.4 Aerial view of the community

Fig. 4.5 Cantagalo in 1950

homes on the hillside mainly because of the proximity to workplaces, using their cheap houses as a solution for the lack of financial resources (Fig. 4.5).

In the 1960s it was declared that the construction of brick houses in the community was prohibited in order to make it easier to remove the houses, as it was already been undertaken with the houses in the favela of Catacumba located near Lagoa Rodrigo de Freitas. Although the mayor of the city, Carlos Lacerda, proposed to remove all slum dwellers from the South Zone of Rio, there was an opposing action, because many of these residents, instead of leaving for distant neighbourhoods, moved to other hills close by.

This territory is divided into 10 sub-areas: Nova Brasília, Buraco Quente, Quebra Armada, Terreirão, Cantagalo, Pavãozinho, Pavão, Serafim, Carangueijo and Vietnam, with a population estimated at 10,338 inhabitants according to data provided from the municipality in 2014 (EMOP 2014).

The issue of forced removals lead a strong social reaction from residents, students and intellectuals, who are committed to start a democratic process of urbanisation in the favelas. Due to the high costs of removal of families and the strong resistance from the population and organized groups, a reformulation of strategic urban planning occurred in the late 1970s, which sought to broaden the dialogue and participation in the administration.

In the 1980s, under the government of Leonel Brizola, a process of transformation in the politics of urbanisation in the slums began. The main idea was to offer land tenure for the residents from low-income communities through land regularisation in a program called Every Family a Lot. In addition to the regularisation, new infrastructure was proposed, which conflicts with the slum removal process. As a result, the permanence and disorderly growth of the communities in question led to the need to install more infrastructure, such as sewage, sanitation, water supply, electricity, basic health services and education, among others. The housing policy changed, with a paradigm shift. The issue of removals was replaced by Every Family a Lot Program, which emphasized land regularisation, urban improvements, maintenance of the population in the community itself, among others. However, despite the agreements between the various public agencies, it cannot be said that there was integration of the powers that operated in the favelas. One of the first communities chosen by the State Government for this program was Pavão-Pavãozinho-Cantagalo.

4.4 The PAC Programme in Pavão-Pavãozinho-Cantagalo

The main idea of the PAC for the community was to redefine the urban layout of the area and to build new housing units. These actions, coupled with the presence of the Pacifying Police Unit (UPP) and the construction of new accesses, such as the elevator connected to the subway system, have created new social opportunities. However, basic inequities prevail, which make social inequality visible just in front of the inhabitants of the "city", especially in its immediate surroundings such as Copacabana, Ipanema and Lagoa districts.

The interventions of the PAC Pavão-Pavãozinho and Cantagalo (Fig. 4.6) began in 2007 and were carried out by the construction company called OAS. The total investment was about US$ 26.5 million. PAC Pavão-Pavãozinho and Cantagalo comprised of infrastructure works (water, sewage and drainage), the construction of

Fig. 4.6 Aerial image of the community

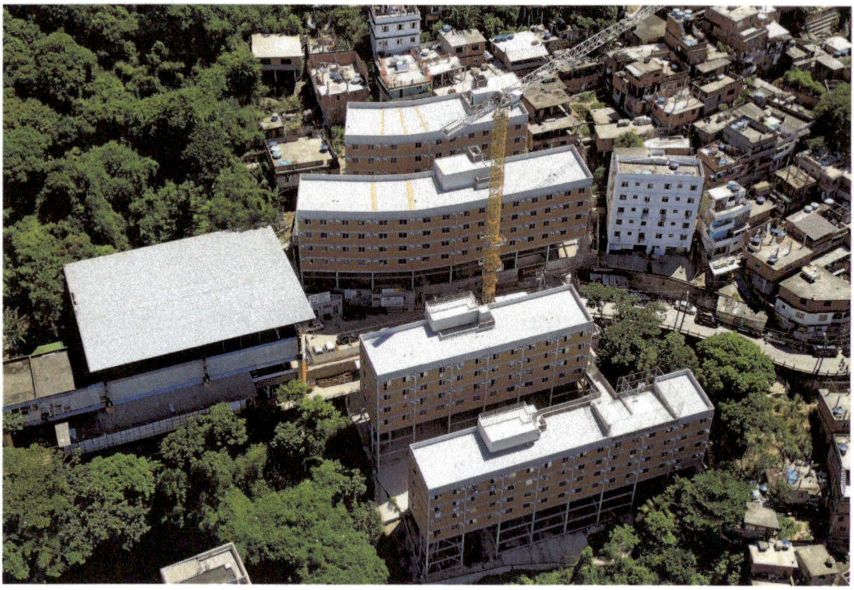

Fig. 4.7 Building apartments—PAC 1

120 housing units in Cantagalo (Fig. 4.7) for families that were living next to a building site and had to move, for people that were removed from a building, which was needed for the installation of a UPP' and the construction of "longitudinal alleys" in both communities, aiming to improve the access for residents and services. The works were interrupted in July 2010, when resources dried up.

In November 2011, the second phase of PAC Pavão-Pavãozinho and Cantagalo began (Fig. 4.8). For this phase, the construction of two vehicular routes was planned, with the extension of Pavãozinho in Pavão-Pavãozinho, and Avenue and Custódio Mesquita in Cantagalo. The project also planned for the construction of an

Fig. 4.8 Men at work

Olympic Village at the top of the community in the form of 76 new housing units on Saint Roman Street and the renovation of a building called "Casarão" that was in a precarious state with 27 apartments and is located in Pavãozinho.

The project to build an Olympic village in the upper part of the community, known as Carangueijo, was denied by the City Hall. The land was considered as inappropriate after evaluation of the risks of the area, and being located in an area of environmental preservation (above 100 meters high). To date twenty-four families remain in 'social rental' receiving aid of US$ 200,00/monthly, until new housing units are ready.

Between 2011 and 2013, the construction of 16 housing units (Fig. 4.9) was completed and the construction of Avenue Custódio Mesquita started, in addition to the renovation of the Casarão.

Fig. 4.9 The housing units at Pavão—PAC 2

Fig. 4.10 Cultural exhibition in the subway

4.5 The Actors and Their Contribution

The urbanisation process was conducted in a participatory way with engineers, architects, urban planners and social workers, residents, leaders, NGOs and Universities. The team of EMOP (State Company of Public Works) was responsible for the supervision of projects and Social Work. Residents, NGOs and leaders accommodated the monitoring works and relocation of residents with ample debate. They also participated proactively in the definition of social projects to generate income for the residents. There were several Management Committees created to monitor the whole process. A multidisciplinary team of about 30 professionals from architecture areas, urban planning, social area, geography and others, took care of the Social Work.

During the PAC Program the government chose to integrate the social issue, urbanisation works and land regularisation, in order to guarantee a permanent social dialogue and to improve the quality of life of this population in an effective and permanent way.

In each of the territories points of service to the population as a space for permanent dialogue were set up, wherever the works in those places impacted the population. It was agreed to call these spaces 'social offices'. Such spaces have become the physical basis for the operationalisation of socio-technical interventions.

The based on three different areas of activity:

1. Home and Business Census

   The purpose of the household census was to create a more veracious picture of the communities involved and to provide subsidies for all the work to be done. The result is a complete analysis of income, schooling, attendance and local services, condition and quality of life. These data can be easily compared with the data obtained by the National Census.

   The business census has identified and evaluated the entirety of formal and informal enterprises as well as the entrepreneurs within these communities.

   The realisation of the census in the favelas was a pioneer project and represented an important progress in the census survey model that had been carried out in Brazil. The methodology of the work with its territorial and temporal permanence was an important milestone that presented a more adequate approximation of the quantity of the population and its living conditions. This instrument allowed the federal, state and municipal governments to propose projects and public policies that are specific and appropriate for these areas.
2. Management of Impacts and Territorial Development

   2.1 Impact Management: Large physical interventions impose several risks to the enterprises and social issues could make the ambitions difficult to achieve. That is why 'Impact Management' aims to minimise the negative impacts caused by works, such as the disturbance to mobility, the interruption of services and the relocation of residents. It also maximises the positive impacts, such as the generation of jobs and income, the improvements in infrastructure and services for the citizens and the territory, and the expansion of the social dialogue with local families and companies, which have to be relocated in order to look for solutions that are able to meet the needs of everyone in the area.

2.2 Territorial Development is based on the sustainability threesome, whose elements are economic growth, environmental balance and social equity. The search for the required conditions for the territorial and sustainable development of the community takes place through the following stages:

   i. Mobilisation and organisation;
   ii. Shared management;
   iii. Local knowledge and diagnosis;
   iv. Elaboration of the Future Vision Plan;
   v. Economic growth;
   vi. Social development;
   vii. Resource sustainability;
   viii. Democratic management.

3. Land regularisation arranges for every family the title of the real right over the soil and/or construction in which it lives. This certificate will be given through the instruments of use, which normally is used for the private domain, and grant of real right of use for the areas of public domain, enabling the families the security of tenure. This process has resulted in the Complexo do Alemão regularise 18,000 residences, in the favela of Rocinha, 5000 residences, in Manguinhos 4200 residences and in Pavao Pavaozinho and Cantagalo 4700 residences.

In the aim is always to develop the territory in a sustainable and prospective way, and in this sense the logic of Participative Urbanism, has always been taken into account in the program. It seeks always to minimise the negative impacts on the daily life of the population, and to generate job opportunities and income through hiring labour for the works in the areas from the area itself.

Within the pillar of Social Work, committees were formed for the Follow-up of Works, the Reorganisation and Generation of Income. These committees were responsible for the publication of all actions and activities, to minimize conflicts arising from the works and to establish a close relationship with the government.

Another important point was, in addition to the implementation of the Integration Plan for Work and Income, to support families/businesses to adapt to heritage conservation and condominium management of their post-occupation estates, in order to guarantee the sustainability of these families after having undergone transformations in the their lives or in their business.

In the proposed participatory methodology, several meetings and forums were held with the participation of local residents and NGOs from the three areas. The main goal of those meetings was to debate all the phases of the PAC interventions with the residents and to receive contributions from them about this slum upgrading project. One of the principal issues were about the relocation process of residents and commercial areas. They were worried about this situation and most of them wanted to stay living in the community and not to be relocated to other areas.

Another important step in the developed methodology is the health and environmental education axis. From this axis educational and awareness campaigns were developed to minimize negative impacts on the environment. This is one of the most relevant and difficult aspects to tackle. It should be emphasized that in general the population itself handled the solid waste management, since the Public Authorities did not carry out regular collection in these region areas, and in some cases did not even have a garbage collection routine due to logistics difficulties and did not have vehicles suitable for collection in circulation. At other times, territorial control by drug traffickers prevented public urban cleaning work.

Because of all this, the residents of those communities were not accustomed to have their garbage collected on regular days, which was introduced through the PAC interventions. To carry out educational campaigns, workshops for reuse of recyclable material and establish an effective partnership with the municipal waste collection company are elements of a successful environmental strategy.

Part of the environmental policy, ecological trails (Fig. 4.11), cultural exhibitions (Fig. 4.10) and gastronomy fairs (Fig. 4.12) were designed to create a greater awareness of the environment in the population (Fig. 4.12).

Fig. 4.11 Inauguration of the ecological trail

Fig. 4.12 Gastronomy fair

4.6 Conclusion

The slum upgrading process, which was carried out as part of the Growth Acceleration Program (PAC), had a strong impact on several communities, such as Pavão-Pavãozinho and Cantagalo. For about seven years, important infrastructure projects, the construction of public facilities and new housing for low-income families, have been completed in this community. All this work and effort was a great success because of the guarantee the participation of the community can follow and discuss all the phases of the work, avoiding unnecessary opposition and enlarging the sense of belonging as result of the clarity and transparency in the debates. Another important aspect in the conduct of the process was the Health and Environmental Education axis, where residents could be made aware of the importance of the environment through actions related to garbage collection, environmental sanitation and also the preservation of public spaces and the improvement of the quality of life. In this sense, the community project demonstrates how important environmental initiatives are for changing a culture that combines environmental concern with a healthy food project.

References

EMOP. (2014). Termo de Referência (ToR)-Pavão Pavãozinho e Cantagalo/ complemento. Plano de trabalho Técnico Social em apoio às obras. EMOP—Rio de Janeiro.

IBGE. (2010). URL: http://seriesestatisticas.ibge.gov.br/series.aspx?vcodigo=POP122, Accessed: 17 November 2016

Ministério das Cidades. (2013). Brasil Conhecimento e planejamento integrados: trabalho social em programas e projetos de habitação de interesse social: Manual Temático Volume 1. Aliança das Cidades. Brasília, secretaria nacional de habitação.

Author Biography

Ruth Jurberg Urban Planner, working at the State Government of Rio de Janeiro since 1983. General Coordinator of PAC Program since 2008.

Chapter 5
Involvement of Residents

Rob Roggema and Ruth Jurberg

5.1 Introduction

The implementation of a place for growing food in the favela is not easy. The residents are used to obtaining their food elsewhere, existing shops are not always happy with people starting to grow their own food, and safety of the once created placed stays an issue. Anyone who thinks this kind of projects can be realised in a top-down way must be confused. The only way top organise support for changes in the direct urban environment requires a participative design process. Not in the last place because most of the spaces where food might be grown is on top of roofs in privately owned houses. Therefore the way to involve residents is important. Not only in case of food planning, but in every change in people's direct environment it is important to gain support of the people and allow them to co-design. In this chapter two examples of public involvement are presented, the design charrette and the Imaginarium Workshop, and the way residents were involved in the first FoodRoof will be described.

5.2 Food Planning as a Process

Understanding the chains of food production, applying the principles of metabolism and plan at the appropriate scale is important but is not successful without an inclusive planning process. The question is however, at which scale and with whom

R. Roggema (✉)
Faculty of Design, Architecture and Building, University of Technology Sydney, Ultimo, NSW, Australia
e-mail: rob@cittaideale.eu

R. Jurberg
State Government of Rio de Janeiro, Rio de Janeiro, Brazil

R. Roggema (ed.), *Food Roofs of Rio de Janeiro*,
DOI 10.1007/978-3-319-56739-6_5

this process should take place. The food planning process could become a platform for all actors in their area, in this case in Cantagalo. This platform is a planning tool facilitating and enhancing a multi-stakeholder and cross-disciplinary planning process, and addresses uncertainty as part of a design process in a dynamic world. Such a panning process is not linear from analysis to implementation, but works cyclically and iteratively, jumping back and forth between steps. During the planning process other techniques and tools can be used, such as multi-stakeholder consultations (Keskitalo 2004), design charrettes (Roggema 2013), data mining from crowdsourcing (Heipke 2010; Elwood et al. 2012), 30–30 exercise (Roggema et al. 2013), backtracking (Roggema 2009), prototyping (Budde et al. 1992; Buchenau and Fulton 2000), or foresight (Destatte 2010). The common and essential element in these tools is the capacity to enable creativity. Food planning is a process, and not a blueprint to execute in a certain area.

5.3 Design Charrettes

5.3.1 Definition

In dealing with complex design problems the design charrette is very suitable method. Charrettes are successfully used in controversial and complicated design and planning problems. They are used in (Lennertz and Lutzenhiser 2006):

- High stakes projects involving substantial public and private investment;
- Volatile yet workable political environments—situations that are “hot” but manageable;
- Complex design problems;
- Real projects that include imminent development.

The charrette is defined as: ‘a collaborative design and planning workshop that occurs over four to seven consecutive days, is held on-site and includes all affected stakeholders at critical decision-making points’ (Lennertz and Lutzenhiser 2006) and ‘a time-limited, multiparty design event organised to generate a collaborative produced plan for a sustainable community’ (Condon 2008).

In general, the design charrette consists of three phases: preparation, visioning and implementation (Roggema et al. 2011). The core interactive design process takes place in the visioning charrette. The group of participants ideally consists of a mix of age-groups, backgrounds, professions and gender. A typical visioning charrette program starts with introductions on the theme, task and urgency, followed by an exercise to make people conscious of time frames. In the next phase, participants work in mixed groups on design sketches at different spatial scales. This is an iterative process, in which people can be regrouped and the designing at different scales occurs. The design charrette concludes with a tactile session in which 3D models are built. People stay in the same group in order to raise competitiveness. The results are presented to a decision-making collective. This causes a mechanism of the need to perform and present the best ideas as a group.

The preparation consists of the contacts with potential participants, explaining the purpose and the atmosphere of the charrette, (re-)formulation of the design task, the compilation of the program and organisational aspects, such as the venue and other means.

During the implementation charrette the same participants work on reaching consensus about the design and formulate projects, which have the support of the group and are essential to realise the strategy (the so-called catalyst-projects).

5.3.2 *Example: Binder-Sum*

In Binder-sum, a small regional community in North-eastern Mongolia a design charrette has been held in August 2015. The aim here was to make a design for developing the town as well as the touristic potential in the region where, according to the story, Genghis Kahn was born (Onon 2001). In the area, nomadic farming is traditionally important, but recent developments place the town under threat of economic downturn and depopulation.

The participants in this charrette were mainly local citizens, such as farmers, entrepreneurs and older people, government representatives and local and regional politicians, designers and experts from Ulaan Bataar and abroad.

The program started with several days of site visits. After this, the group gained insights in the history of the area and defined desired future prospects in the 30–30 exercise, and presented the outcomes to each other. This was followed by designing a plan for the future of the area at the regional scale, including an intermediate presentation to the other subgroups. During the final part of the charrette the participants build their ideal future in 3D using plasticine and presented the final results to a group of decision-makers.

The results of the Binder-sum charrette consisted of several designs for a diverse range of touristic developments in the Binder region, such as ideas for new local tourist settlements, including farm stays with local families, new hiking and horse trails with camps, locations for fishing and boating adventures, creating viewpoints on mountain tops, sustainable resorts in the steppe landscape and mountains and the restoration and accessibility of cultural monuments, such as the birthplace of Genghis Kahn, the oldest known parliament in the world and reminiscents of the Turkish and Mongol cultures, such as walls, inscriptions and stone carvings. The touristic plans were accompanied with efforts to make the area more sustainable, such as regeneration and replanting of forests, capturing and storage of water, ecological development, and the building of new local and sustainable houses.

5.3.3 Benefits of Design Charrettes

Working with design charrettes to create both spatial as social urban resilience has multiple benefits:

1. The complex design tasks are, during the charrette process, undertaken in an interactive and trans-disciplinary way, with people from different backgrounds who work together and come to innovative, never thought of, propositions. It makes the driving forces of participants' desires manifest and open for discussion.
2. The tactile methods used during the design charrettes appeal to the right brain-half, which allows people to use their creativity, dive in their imagination and think of suggestive ideas. This way they can leave their rationality behind and aren't bothered by their organizational interests. Instead they are open to negotiations and seek mutual benefits. Mapping is used to make this visible and is a means for an interactive way of exchanging convictions, insights and desires.
3. The results of design charrettes are conceptual plans and designs with a series of concrete, spatial propositions, which anticipate future uncertainties. These propositions may incorporate sudden climate events, such as a flood, drought or fire, but are also able to deal with an economic downturn. The set of design proposals together creates a vision for a resilient environment, which is capable of dealing with sudden changes.
4. A concrete result is a drawing or 3D-model of the shared visions for a desired future. This end-result is supported by the participants and can be used as the basis for further work, for instance to program realisation projects or to base new policies on. The drawing counts for the participants of the design charrette as a contract, which cannot be changed without accordance of all involved.
5. In design charrettes an atmosphere is created to facilitate the conversation between different participating groups, such as citizens, academics, policy-makers and entrepreneurs. This makes it possible for the people who take part to regain trust in governmental planning and confidence in the benefits of design in general.
6. People also regain trust in politics, because the design charrette process takes the participants as serious contributors to policy making and planning, hence people feel they are listened to.
7. The process to reach the end result is relatively cheap. In one week, or twice a three-day session, a complete Master Plan for a resilient urban environment or landscape can be developed. This Master Plan is built on the expertise of local experts, reason why local communities will support the content. A charrette design process requires a good preparation, but process is relatively quick, with low resistance. The charrette feeds the design with new insights and coherent innovative spatial proposals, and this makes decision-making easier.

5.4 The Imaginarium Workshop

The Social Work is a component of the PAC Program and aims, on the one hand, to minimize the negative impacts of the physical interventions carried out by the state in favelas and settlements and, on the other hand, to create development strategies for the territory.

The so-called Imaginarium Workshops were an integral part of the Technical and Social Work Projects, elaborated for the communities of Complexo do Alemão, Rocinha, Manguinhos, Tijuca, Jacarezinho, Mangueira, Lins, Jacarezinho, Cidade de Deus among others. The highest priority was the participation and involvement of the residents in the PAC. The proposal was to involve the community in the project and to have as a final result the partnership of all through the care of the public spaces and the maintenance of the panels made from the drawing and graphic expression made by the residents themselves.

The main objective of the Workshops of the Imaginary (Imaginarium Workshop, Fig. 5.1) was to creatively harvest the memory and identity of the areas that had their landscape altered due to physical interventions. These memories and identities were harvested through letting the population and children draw their imaginations (Figs. 5.2 and 5.3).

Fig. 5.1 Imaginarium workshop

Fig. 5.2 Children's design

The final objective was to have drawings capable of illustrating, for example, the cable car stations and the 4th Street in Rocinha with the characteristic and identity of the residents living there. This way the residents are empowered in the mosaic technique, making the teaching of a technique possible that can be used as income generator later. Large mosaic panels on 4th Street and cable car stations were set up. The illustrations on each of the panels show the identity of each of the communities. This way a greater proximity between the equipment that had to be realised in the favelas and the residents was created, which increased the feeling of belonging.

The methodology is to diagnose, through drawings, the wishes of the community about the place where they live. The designs that best portrayed these desires were chosen to represent the community, and the project made it possible to transform these designs into mosaic panels, executed by the residents, participating in the workshop, themselves. The purpose of this method, which was applied in two territories, was to get hold on the history and identity of the communities in a playful way, while they underwent the realisation of equipment built in these places.

As protagonists of a growing process of popular participation in the communities of Rio de Janeiro, the residents were invited through posters and invitations in the areas where the workshops would be held. The number of participants varied greatly, but in total, there were approximately 250 residents participating.

Fig. 5.3 Design of population for Palmeira station

During these workshops the residents were guided on the works that were going to be realised in the community. The people were then invited to express themselves with drawings on their desires for the area. To hold the workshop, residents of all ages were invited and were grouped together (Fig. 5.4) to make drawings (Fig. 5.5) and then present their drawings. These drawings are later transformed into designs by teams of architects and mosaic art panels such as those placed in cable car stations.

In the workshop held in the Complexo do Alemão, the communities chose the names of the cable car stations as subject of their designs. Today, these designs form the ornaments of each station in mosaic panels, such as for Palmeira station (Fig. 5.6). In Rocinha, the panels chosen and executed by the community are distributed at several points on Rua 4 (Figs. 5.7 and 5.8).

The Social Technical Work team considers it fundamental that the residents are involved throughout the entire intervention process, including the humanisation of the newly built spaces. With this it is ensured that the new "landscape" is incorporated into the daily life and cared for by the residents. It is worth noting that up to now the panels have not been damaged by vandalism, and the respect and

Fig. 5.4 Working in groups

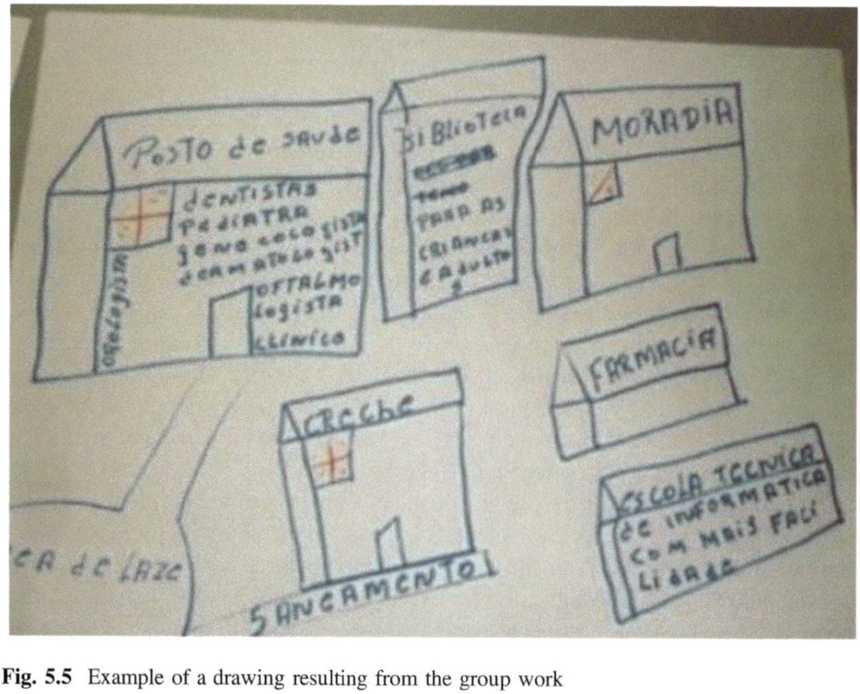

Fig. 5.5 Example of a drawing resulting from the group work

Fig. 5.6 Mosaic at Palmeira station

Fig. 5.7 Rocinha panel predio

admiration that the residents have for seeing part of their stories on large murals exposed to the world is remarkable.

The Imaginarium Workshop technique could be replicated in many different circumstances when residents are to be involved in a participatory process. In this

Fig. 5.8 Rocinha mosaic

specific case it helped to create identification of the community with what had been built by the PAC. The approach can be replicated in any other circumstance and with little resources. It is imperative that the workshop takes place on the spot, so the residents can access the venue easily. It can even be organised in an outdoor square.

5.5 Involvement in FoodRoof

The design process of the FoodRoof was not organised as a design charrette, nor has it much similarities with the square project. However, the residents were involved in the design process in different ways.

Rafael Levinho was our main partner in the initial phases of the project. He is a key person in the Team VB jiu-jitsu club and was one of the first with interest in having a FoodRoof. We visited his house many times and came up with new design propositions all the time. He responded and influenced the way the roof could look like and could be used. He was happy with the design but in the end time constraints prevented the FoodRoof been realised on his roof.

The EMOP helped us in getting contact with resident groups and their representatives. This way we could obtain contacts in the favela, and also could gain support for the initial ideas of the FoodRoof. After the presidents of the community groups supported the FoodRoof concept it was much easier to find individual owners who might be interested in the realisation of a FoodRoof.

Along these lines we also met Marcelo Assunção, and he cooperated very well with the design and building team. He supported the realisation of the FoodRoof during the building process and he could influence some of the solutions in detail, even during the building period. Additionally, he was a mouthpiece for other residents. Through him many others in Cantagalo understood what the purpose is and which benefits of the FoodRoof could offer.

5.6 Conclusion

As the examples in this chapter show there are many different ways of involving residents in a design process. There is no single way of doing it right. The most important element of a participatory design process is the way people are treated. The residents need to be taken seriously as a source of local knowledge, caring for their own environment and the ability to invest time and attention in their neighbourhood. Any form of top-down planning or the impression of it makes this subtle balance vulnerable. As soon as people feel there is a hidden game being played, or the suggestion that it is, they will object the plans proposed. But as soon as the approach is genuine, they will easily collaborate and contribute to the quality of the designs.

For slum areas this is very important, as the people living there are generally not the people that love the government or other authorities. They should be treated as full citizens that are taken seriously and are listened to. Trust is the key here. As soon as people trust each other, beautiful plans are possible. Moreover, the people will guard the outcomes of these design processes and maintain custodianship over it.

References

Buchenau, M., & Fulton, J. (2000). Experience prototyping. In *DIS'00 Proceedings of the 3rd Conference on Designing Interactive Systems: Processes, Practices, Methods* (pp. 424–433). New York: ACM. doi:10.1145/347642.347802.

Budde, R., Kautz, K. H., Kuhlenkamp, K., & Züllighoven, H. (1992). Prototyping. In R. Budde, K. H. Kautz, K. Kuhlenkamp, & H. Züllighoven (Eds.), *Prototyping an approach to evolutionary system development* (pp. 33–46). Dordrecht, Heidelberg, London: Springer.

Condon, P. M. (2008). *Design charrettes for sustainable communities*. Washington, Covelo, London: Island Press.

Destatte, P. (2010). Foresight: A major tool in tackling sustainable development. *Technological Forecasting and Social Change, 77*(9), 1575–1587.

Elwood, S., Goodchild, M. F., & Sui, D. Z. (2012). Researching volunteered geographic information: Spatial data, geographic research, and new social practice. *Annals of the Association of American Geographers, 102*(3), 571–590. doi:10.1080/00045608.2011.595657.

Heipke, C. (2010). Crowdsourcing geospatial data. *ISPRS Journal of Photogrammetry and Remote Sensing, 65*(6), 550–557.
Keskitalo, E. C. H. (2004). A framework for multi-level stakeholder studies in response to global change. *Local Environment, 9*(5), 425–435.
Lennertz, B., & Lutzenhiser, A. (2006). *The charrette handbook. The essential guide for accelerated collaborative community planning*. Chicago: The American Planning Association.
Onon, U. (2001). *The secret history of the Mongols*. London, New York: RoutledgeCurzon Press.
Roggema, R. (2009). *DESIGN: Final report hotspot climate proof Groningen* (p. 159). Groningen/Wageningen/Amsterdam: Province of Groningen and Climate Changes Spatial Planning.
Roggema, R. (2013). *The design charrette: Ways to envision sustainable futures* (p. 335). Dordrecht, Heidelberg, London: Springer.
Roggema, R., Horne, R., & Martin, J. (2011). Sharing the climate adaptive dream. In *Proceedings ANZSRAI Conference*, December 2011, Auckland.
Roggema, R., Martin, J., Remnant, M., Alday, G., & Mansfield, P. (2013). Design charrettes in two days: Sea Lake and Bendigo. In R. Roggema (Ed.), *The design charrette: Ways to envision sustainable futures* (pp. 117–149). Dordrecht, Heidelberg, London: Springer.

Author Biography

Ruth Jurberg Urban Planner, working at the State Government of Rio de Janeiro since 1983. General Coordinator of PAC Program since 2008.

Chapter 6
Development of an Overall Design Strategy

Rob Roggema

6.1 Introduction

The first issues that come to your mind when talking about the favelas of Brazil are crime, violence and fear. The reality is different. Walking around Cantagalo favela the tranquillity, safety and friendliness are the first impressions. Despite the low living standards, this place feels like a community and crime seems to be far away (in reality, it is still around the corner). One of the projects that contribute to this atmosphere is the Team VB Jiu-Jitsu school. Up the hill, in an old and underused building every night youngsters come together to train jiu-jitsu, but are also taught lessons for life. Not only physical and technical training is important, but the social values of respect and safety are equally apparent. This has led to a positive change, as the kids forward the message of peace and safety throughout the favela and reach out to their parents, who were probably on the wrong side before but now adhere to the new standards of wellbeing. This projects among many others has resulted in a community with high social values, in which caring for each other, creating a safe place for living and becoming respected citizens are the norm. The only problem seems how to feed the inhabitants.

In this chapter the current programs in favelas are discussed, the spatial context is analysed and design propositions for the entire favela are proposed.

R. Roggema (✉)
Faculty of Design, Architecture and Building, University of Technology Sydney, Ultimo, NSW, Australia
e-mail: rob@cittaideale.eu

R. Roggema (ed.), *Food Roofs of Rio de Janeiro*,
DOI 10.1007/978-3-319-56739-6_6

6.2 Current Programs in Favelas

For a long time favelas have been unsafe, dominated by criminal gangs and drug traffickers. Despite several efforts to create safer and healthier environments the pressure to change life in the favelas could only be build up in the run up to the Olympics in 2016, defined by some as an 'Olympic makeover for the favelas' (Downie 2011). This makeover consists generally of executing infrastructural measures first, such as electricity, light, tele-communication and home appliances, to be followed by health, education, water, transport and employment. This program, initiated by the State of Rio de Janeiro through the Program for the Acceleration of Growth (PAC in Portuguese), started in 2008 with intervening in five favela's, amongst which Cantagalo. The main objective was to 'transform physical interventions into sustainable development, bringing together projects that generate transformation of the territory through economic growth and community participation based on constant dialogue' (Jurberg 2013). This process of pacification was made possible by instalment of the so-called UPP's (Pacifying Police Units), heavy armed police posts at strategic points in the favelas. These policies have lead to both enhanced urbanization and sustainable development (Alvarado 2012).

Many of the projects that aim for social improvements also take into account art. They aim to create employment and income in the favelas by branding them (Melo 2013). 'City officials hope the answer is a program called Morar Carioca, a comprehensive program to bring the settlements fully into the city by extending public services, improving individual homes, and constructing new housing and infrastructure, such as pedestrian passageways and, where feasible, streets that will allow sanitation trucks to pass through the rehabbed neighbourhoods' (Benfield 2013).

Until now, Rio has grown and changed rather haphazardly, without an overarching vision: there's no overall sense of the flow of the city in the second-term strategic plan (Rio Prefeitura 2013), and there is no vision on its physical form (RioReal 2013). Many of the urban design interventions of the past are incongruently focusing on spatial interventions necessary for infrastructure adjustments, pure modernist city developments or processes of public involvement. Recent years illuminate a revival of acupuncturist urban design as individual improvements of streets, avenues or squares rather than impacting on a whole-city design (Brandão 2006). The common idea at that stage was to raze favelas, remove them and rebuild them elsewhere, generally farther out. However, in the most recent years the attitude towards favelas began to shift. These areas also could be seen as places with many urban qualities, qualities that are difficult to develop through planning and which urban planners from all four corners of the world today endeavour to stimulate—often retroactively or facing significant cultural or political challenges. These potential urban qualities include (Williamson 2013):

- Affordable housing in central areas.
- Density promoting and enabling quality of public services without excessive verticality that leads to isolation.
- Pedestrian-oriented planning, encouraging better opportunities for community development and exchange.
- High use of bicycles and public transportation, which have a positive environmental impact on the local and global scales.
- Mixed use (residential over commercial lots) which reduces the need for transportation and stimulates community exchange.
- Living near work, reducing expenses and time on transportation, as well as avoiding overloaded transit networks.
- Organic, or slow, architecture—iterative architecture that slowly evolves adapting to the needs and conditions of the residents.
- High degree of collective action, which not only strengthens community bonds through mutual support, but it also offers economies (savings) with regard to a number of services and materials exchanged or offered in kind.
- Intricate solidarity networks.
- Advanced degree of cultural production.
- Encouraged entrepreneurship, enabled by a constant exchange between residents, the possibility of creating businesses at home and the flexibility made possible by a historic lack of regulation.

These developments do not contain issues such as the demand for food, food security or food production very prominently. Food is not a large part of these programs and projects, with one exception, the 'Hortas Cariocas' project developed by the City of Rio de Janeiro. This gap in policy, programs and projects can also be seen as an opportunity to not only bringing in facilities, such as electricity or the water provision, but to create additional values, which increasing the productivity, social connectivity and economic independency.

6.3 Problem Definition and Objective

In Cantagalo/Pavão-Pavãozinho a total of approximately 20,000 people live. It used to be a very unsafe place, where drug trafficking determined the power relations and criminality was widespread. The residents generally don't eat very healthy and there is not a lot of space for growing food. The strongly constructed houses are closely built together and the sparsely public space mainly consists of small alleys. When it rains, it rains extremely hard and the water runs off the steep hill, discharging not only water but also all the rubbish down the stream. The favela is built on the steep slopes of an enormous rock (Fig. 6.1), which doesn't provide a very fertile underground for growing food.

Fig. 6.1 Cantagalo, built upon a steep slope (*photo* Rob Roggema)

The current food system does not reach the residents of favelas. There are several reasons for this. First of all, the abundance of food in Brazil is available through restaurants, supermarkets or open markets, mainly frequented by the upper-class society. These places are located outside and sometimes far away from the favelas. The effect of this is that there is no fresh food available in the favela. The little shops in the favela only sell pre-packed food, cakes, candy, and beer and cachaça. The physical, economical and mental distance is hard to overcome for the people living in Cantagalo, because the time and money required to obtain food is normally used to earn money. Secondly, the available food, if offered close to the favela, in supermarkets and restaurants, is expensive. Especially around Cantagalo/Pavão-Pavãozinho, the neighbourhoods of Ipanema and Copacabana are very expensive. This leaves the residents of these favelas no other choice than eating the cheapest food around. In general the cheapest food is not the healthiest and freshest food, should there be enough in the first place.

Besides the problematic food supply system, there are two other problems, which together determine the bandwidth of possible improvements.

Firstly, the urban water system is not adequate and cannot provide a sufficient and guaranteed supply to grow food. There is too much water during (heavy) rain, and too little during dry, humid periods. At the source of the water system, rainfall, the available water is of good quality, but this water is transported down the hill through a system of concrete gutters. This water is provided to the neighbourhood

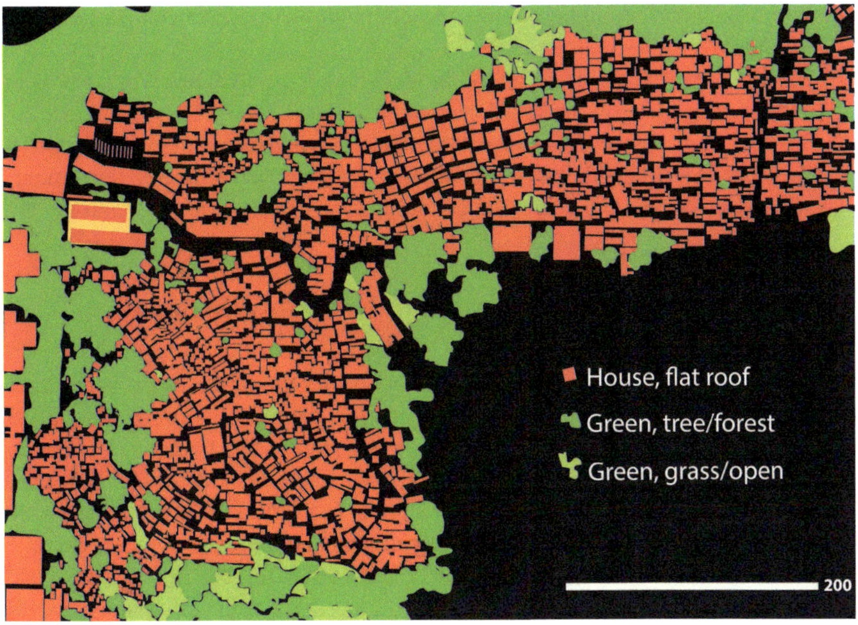

Fig. 6.2 Confined urban spaces in Cantagalo

by the water utilisation company and is pumped up to the upper parts of the area and cleaned before it is distributed. However, the quality of this water decreases seriously when it is transported further downwards, and garbage and dirty water is added to this system. Rainwater is not directly used and currently seen as a burden, which needs to be discharged as quickly as possible.

The other additional problem is the size and availability of open space. In the favela these spaces are confined, very small and urban (Fig. 6.2), with pavements mainly made of stone or concrete. The task for planning and design is to develop smart solutions for a food system that fits in limited space.

The ambition of the FoodRoof project is to develop a local food growing system that delivers fresh and healthy food for the people in Cantagalo (Roggema et al. 2014). Given the circumstances in the community, this system should be simple and safe to operate, must be easy to construct and cheap.

Because of the dense and complicated open spaces in Cantagalo, the best way to implement such a system is to use the roofs to grow food and store water. A lightweight and sustainable system is therefore essential. An aquaponic system, which closes cycles of nutrients, water and energy, and is lightweight, which consists of fish breeding, water storage and growing of fresh vegetable and/or fruit is seen as the most suitable for this context (Roggema 2014c).

To realise this ambition a planning approach needs to consist of the following three dimensions:

1. Work with the strengths within the community to create innovative design solutions;
2. Use the opportunities from a food production perspective, use what is possible, create spaces to grow food at local, small scales;
3. Support start-ups, e.g. interested owners (the owners with wow factor: the 'wowners') and help them with rebuilding/adjusting their houses to implement an aquaponic system and create the conditions for successful agricultural system.

6.4 Framework for the Design of Urban Agriculture

In order to comprehend the different aspects of designing for urban agriculture a framework (Roggema 2014a, b) has been developed (Fig. 6.3). The Spatial Sandwich, described in chapter one, at the right side of this framework is the core driver to link and integrate design decisions to create productive urban landscapes. The framework describes how design and urban agriculture relate, and makes clear how to design at the regional, city, neighbourhood and building level. The left hand side of the framework presents the functionality of the Urban Agriculture system through factors such as population (determining the amount of healthy and collectively produced food), products (crop types, productivity) and economics

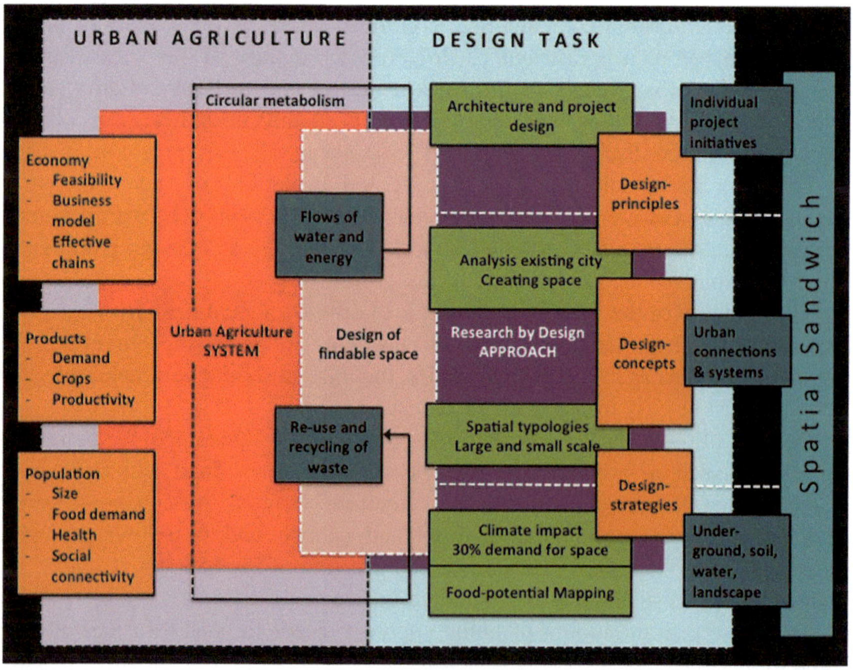

Fig. 6.3 Urban agriculture framework (Roggema 2014a, b)

(feasibility, food chains, business models). These three basic factors interfere with, but also profit from closing and reusing water-, energy-, nutrient- and material-cycles.

The right half of the framework represents Research-by-Design. 'Research by design generates knowledge and understanding by researching the effects of changing the design solutions and/or the context at the same time' (De Jong and Van der Voort 2005). This part of the framework is subdivided in three levels of a 'Spatial Sandwich', connecting individual projects (the Top), urban connections and systems (the 'In-Between') and the underlying landscape (the Basis), see also chapter one. These levels are not necessarily spatial scales, but coherent groups of elements forming a spatial whole.

The two sides of the framework come together in the centre, where they form a platform for the design of space for a productive city. The Spatial Sandwich model has been used to design the FoodRoof for Cantagalo in Rio de Janeiro.

6.4.1 *The Basis (Underlying Landscape)*

Cantagalo favela is located just north of Ipanema and Copacabana beaches. The underlying landscape consists mainly of one big rock with steep slopes (Fig. 6.4) underneath the buildings, which are built in high densities draped over the hill.

Fig. 6.4 Topography of Cantagalo

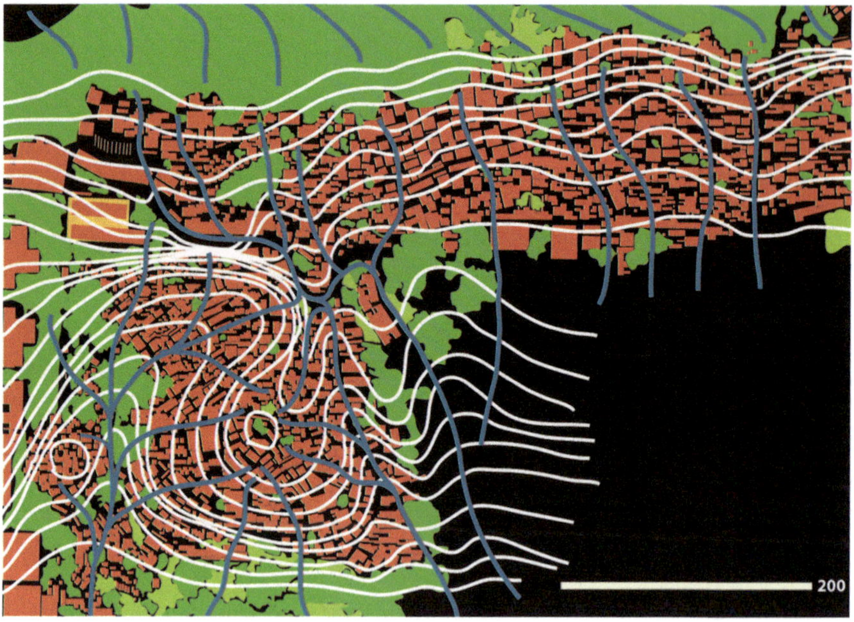

Fig. 6.5 Water discharge in Cantagalo

When it's raining the rainwater is discharged in huge flows due to the slope angles (Fig. 6.5), leaving the area for good.

At the basis level the desired future of the area is determined, resulting from soil (type, fertility), (rain)water and ecology and the landscape, emphasizing the natural or artificial food potentials. A design strategy uses food potentials to create the conditions for growing food. Therefore, in the case of these densely built favelas, the design strategy aims to create a neighbourhood in which as much rainwater as possible is stored and re-used (Fig. 6.6), and fresh food can grow, using artificial soils.

The long-term aim of this strategy is to create enough productive space where local fresh food can be produced, making use of the available stored water. The strategy is threefold:

1. Turn the entire favela into a food producing area; instead of bringing food to the people, the food is grown where the demand is.
2. Use as many roofs as possible, only flat roofs are suitable for productive purposes.
3. Find the available public spaces in the urban environment, such as in open green spaces, or planting edible trees. Green open spaces can be used as veggie gardens.

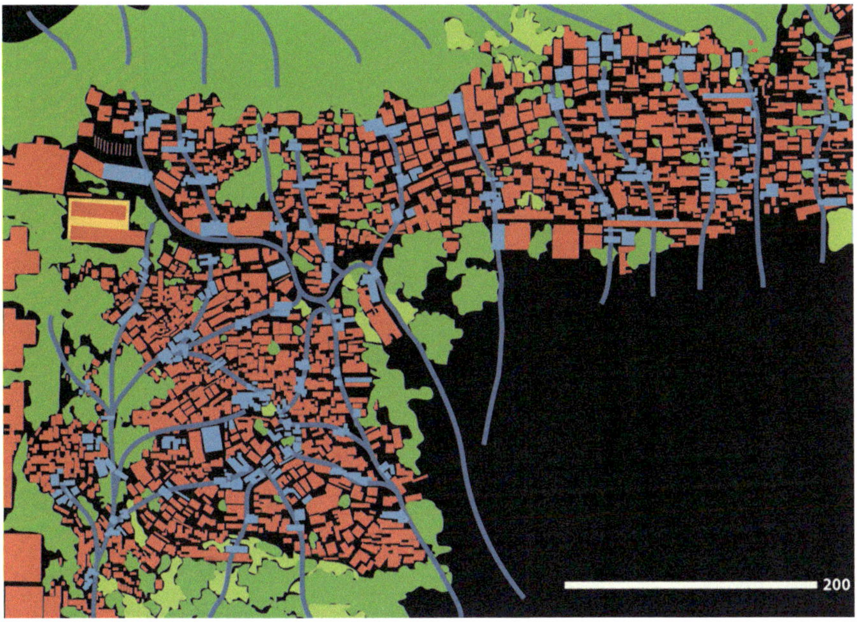

Fig. 6.6 Storage of rainwater on rooftops

Roofs in the favela will be used for different purposes. Some roofs are used for water storage, as a food garden or to generate solar energy respectively, while in public spaces a combination of fruit and leisure functions are preferential.

6.4.2 The 'In-Between'

Conceptual designs provide the conditional framework for elaborative designs and establish linkages between individual projects. Moreover, the food potentials determine how the urban systems are designed and form the basis for concrete designs. These urban systems link closely with the principles of circular urban metabolism.

The existing city is analysed to identify local food-growing places, forming spatial typologies (Point to Point Communicatie 2013), such as place making-, rooftop and high-tech-farming. The specific location of these typologies is determined by the place they can take in the urban systems, such as described in CPULs: 'open landscapes productive in economical and sociological and environmental terms, placed within an urban-scale landscape concept offering the host city a variety of lifestyle advantages and few, if any, unsustainable drawbacks' (Viljoen 2005).

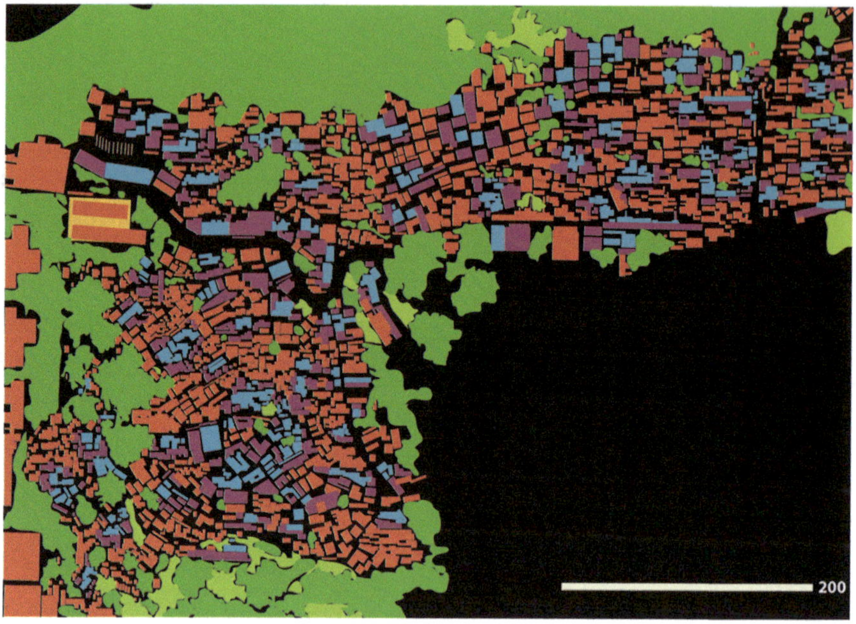

Fig. 6.7 Roofs where rainwater is stored connected to food producing roofs

At the In-Between level a design concept is developed derived from the defined strategy: to use as many flat roofs as possible for water, food and energy harvesting; the ambition is to make use of every available square meter.

The core question is to identify the most suitable places where to store water, where to grow food and where to harvest energy. Together the answers to this question give ratio to a productive rooftop topography. The connections between rainwater storage and the aquaponic food systems are conceptualised in a design proposition for the whole area (Fig. 6.7). When the captured rainwater flows into artificial aquaponic systems on neighbouring rooftops, there is no need for a natural soil, year-round water is guaranteed, the cycles of waste, water, nutrients and energy can be closed, and vegetables, fruits and fish can be grown.

In order to find the most optimal harvesting opportunities specific rules for each of the thematic aspects are defined:

1. Water storage rules:

   - Capture rainwater as early as possible, e.g. at the highest points in the landscape and on top of the highest buildings. The topography of the landscape and buildings determine the locations for early storage.
   - The elevation determines the slope angles. The steeper the slopes the more storage measures are needed.
   - The strength of buildings determines the carrying capacity and whether it is possible to carry the relatively heavy weight of water to be stored.

2. Food production rules:
   - Locate FoodRoofs close to the roofs where water is stored to guarantee availability of water.
   - Make sure the buildings are strong enough to carry the gardens. The aquaponic system is relatively lightweight, which makes many buildings suitable.
   - The roofs must be accessible.
   - Nutrients must be available.
   - The FoodRoofs need to be social acceptable, both for the owner, who initiates and is eager to start the FoodRoof, but also for the neighbours, who need to accept a garden next to their own house.
3. Energy harvesting rules:
   - The orientation of the house or roof, oriented towards the north, northwest and northeast sides.
   - The highest buildings (out of shade), relative to their neighbouring buildings, are most suitable to use.

6.4.3 The Top

At the level of the design principles, the Top level of the Spatial Sandwich, a concrete design for a rooftop proposes four design elements, which are spatially organised according the height of the roof, the sunlight and shade during the year and the space available (Broekhuis and Drissen 2014).

These four elements are the storage of rainwater in fish nursery basins, a hydroponic (horizontal) vegetable garden, a vertical fruit garden and a solar powered water pump to recycle used water back to the fish basins.

At this 'Top' level 'inserted productive urban landscapes' (Viljoen 2005) embedded in the urban context and the underlying landscape, are designed. The design principles need to be linked with the genius loci in the urban system. The way design principles are applied in the designs depends on the aim and context of the design task. The principle distinguishes one design from the other. It creates harmony, or not. It creates balance, or not. And so on.

6.5 Conclusions

Thinking about where Urban Agriculture is needed the most the favelas in Rio de Janeiro make a good chance. Where it sometimes looks like a luxury playing tool and some sophisticated elite movement in western countries, in Brazil it is down to earth needed to supply healthy food to people that have no access to it. The approach presented in this chapter links the qualities of the underlying landscape, with the urban infrastructure systems and the local project design. The connections between the three levels are essential. The local underlying landscape provides hardly any potential to produce food. Therefore, the need to use resources very efficiently is very important. Without this, food production would be impossible. Embedding the FoodRoof in the urban water system and create a production system that closes its cycles provides a sustainable contribution for the residents, which simultaneously gain access to healthy food. The choice for an aquaponic system is not only interesting from a food perspective. It provides a broad range of crops and fish, is able to recycle water and waste, but it is also lightweight and easy to construct. This artificial system of urban agriculture used in the FoodRoof does not require any soil, preventing the necessity of heavy constructions and fitting very well in the rocky basis and steep slopes underneath. The design of the FoodRoof illustrates also the working of the Spatial Sandwich. It is not the productive project on its own, but the impact of the surrounding system on the local project as well as the meaning the single project has as an example for followers. This roof is not on its own, it announces the coming of more and coherent gardens in Cantagalo. Together they will have the capacity to bring sufficient good food to the residents and in particular to the kids of the jiu-jitsu school.

To support the residents of the favelas, food needs to be grown in the vicinity where it is consumed. It requires a system that fits in the urban environment, provides fresh and healthy food, must be simple and safe to operate, and is easy to construct and cheap. Aquaponic systems are lightweight and sustainable with closed cycles of nutrients, water and energy, and consist of fish breeding, water storage and the growth of fresh vegetables and fruit.

There are 3 success factors for designing, planning and realising a FoodRoof:

1. Work with the strengths within the community. They need to be the owners of the project. When the local residents are made co-creators of the project they become committed, are open for collaboration and support and feel more responsible for maintenance.
2. Use local food potentials. In the case of favelas there is not much space, but the roofs are flat and strong enough. There is no soil, so an artificial system is preferential. To design the roofs, the micro-climate, natural shade, humidity, but also the place where rain water can be caught and stored, determine the design.
3. Support start-ups first and then accelerate to extend the project beyond the initial pilot to the entire favela.

References

Alvarado, P. (2012). *The Urbanization of Rio de Janeiro's Slums, a model for sustainable development*. www.treehugger.com/urban-design/the-urbanization-of-rio-de-janeiro-s-slums-a-model-for-sustainable-development.html. Accessed 16 March 2014.

Benfield, K. (2013). *Stabilizing & Greening the Favelas: Rio's Formidable Challenge*. http://sustainablecitiescollective.com/kaidbenfield/188166/stabilizing-greening-favelas-rios-formidable-challenge. Accessed 1 March 2014.

Brandão, Z. (2006). Urban planning in Rio de Janeiro: A critical review of the urban design practice in the twentieth century. *City & Time, 2*(2), 37–50. http://www.ct.ceci-br.org. Accessed 11 March 2014.

Broekhuis, B., & Drissen, M. (2014). *Food growing roof terrace; instruction manual aquaponic system, Roof Terrace Rafael Lezinho*. Velp: Van Hall Larenstein.

De Jong, T. M., & Van der Voordt, D. J. M. (Eds.). (2005). *Ways to study and research urban, Architectural and Technical Design*. Amsterdam: IOP Press BV.

Downie, A. (2011). *Rio gives its favelas a Pre-Olympics makeover*. http://content.time.com/time/world/article/0,8599,2091817,00.html. Accessed 11 March 2014.

Jurberg, R. (2013). Connecting Complexo do Alemão. In *Urban Age City Transformation Conference, London School of Economics and Political Science*, Rio de Janeiro, 25 October 2013.

Melo, D. (2013). *Rio de Janeiro's favelas become a brand*. http://infosurhoy.com/en_GB/articles/saii/features/economy/2013/04/23/feature-02. Accessed 25 February 2014.

Point to Point Communicatie (Red.) (2013) *Stadsboeren in Nederland; Professionalisering van de Stadsgerichte Landbouw*. Rotterdam: Ministerie van EZ, Ministerie van IenM, Van Bergen Kolpa Architecten, LEI, De Volharding Breda, Priva.

RioReal. (2013). *Urban planning in Rio de Janeiro*. http://riorealblog.com/2013/10/04/urban-planning-in-riode-janeiro/. Accessed 17 March 2014.

Rio Prefeitura. (2013). *Pós 2016, O Rio Mas Integrado e Competitivo. Plano Estratégico da Prefeitura do Rio de Janeiro 2009–2012*. Rio de Janeiro: Rio Prefeitura.

Roggema, R. (2014a). Framing urban agriculture: The quest for new design concepts. In *Proceedings ECLAS-conference*, 21–23 September 2014, Porto.

Roggema, R. (2014b). It's time for the McAg: Finding spaces for productive cities in a spatial sandwich. In *Proceedings IFLA 2014*, 4–6 June 2014, Buenos Aires.

Roggema, R. (Ed.). (2014c). *The Nutritious City; The Biospheric Foundation. Lecture Prof. Greg Keeffe. Lecture Series Design for Urban Agriculture, VHL University of Applied Sciences*. Velp: VHL Press.

Roggema, R., Pugliese, A., Drissen, M., & Broekhuis, B. (2014). The FoodRoof: How Cantagalo and Pavão-Pavãozinho favelas grow their own food. In R. Roggema & G. Keeffe (Eds.), *Why we need small cows. Ways to design productive cities* (pp. 207–229). Velp: VHL Publishers.

Viljoen, A. (Ed.) (2005) CPULs: *Continuous Productive Urban Landscapes. Designing urban agriculture for sustainable cities*. Amsterdam, Boston, Heidelberg, London, New York, Oxford, Paris, San Diego, San Francisco, Singapore, Sydney. Tokyo: Architectural Press, Elsevier Ltd.

Williamson, T. (2013). Land tenure and urban planning in Rio de Janeiro's favelas. http://rioonwatch.org/?p=11075. Accessed 16 March 2014.

www.biosphericfoundation.com. Accessed 29 June 201

Chapter 7
Development of the FoodRoof

Rob Roggema

7.1 Introduction

At the scale of the entire globe it can be calculated that we produce enough food to feed nine billion people. Mathematically, this is possibly right. However, many (weak, poor) groups still do not have access to food. This paper reports a way to support these groups to gain access to healthy food. In Rio de Janeiro the residents in the favelas eat candy, potato chips, pre-wrapped cakes, and drink booze and soft-drinks: not the most healthy diet. The 'FoodRoof' is introduced to support local residents to grow their own healthy food. This FoodRoof is a design for an aquaponic food-system on the roof of an individual house. The first FoodRoof has been realised in 2014 in Cantagalo favela and provides fish, vegetables and herbs for the residents of the house beneath. The system closes cycles of nutrients, water and energy and prevents waste to be dumped in the water system and Guanabara Bay, an additional advantage. In this chapter the process how the first FoodRoof could be realised is presented, from the first idea until realisation.

7.2 Problem

Recent literature on food and slums often reports on the vulnerability, malnutrition and lack of food security in slum areas (Build 2010; Faye et al. 2011; Kimani-Murage et al. 2014; Mohan 2015). Strange enough, comments on food in Brazilian favelas are far more positive (Flavors of Brazil 2012; Rothman 2014;

R. Roggema (✉)
Faculty of Design, Architecture and Building, University of Technology Sydney, Ultimo, NSW, Australia
e-mail: rob@cittaideale.eu

R. Roggema (ed.), *Food Roofs of Rio de Janeiro*,
DOI 10.1007/978-3-319-56739-6_7

Barchfield 2013; Petrini 2014), although these articles report more on the entrepreneurship of favela residents than the food they eat themselves.

The current global food system does not reach the people that need it most and it is very doubtful whether a future global food system will. Currently, the shops in Rio's favelas contain candy, chips, wrapped cakes, soda, beers or even stronger drinks. The children, if they go to school at all, eat a bag of potato chips for lunch, and for breakfast, and for dinner. At a very young age the kids start drinking, and follow an, everything but healthy lifestyle.

The ambition of the FoodRoof project is to provide food to these people by giving them direct access to an individual food production system, on their roof. If we can create a food-producing roof, a FoodRoof, which is easy to build, lightweight and can produce nutritious and healthy food for the family living beneath the roof, people in the favelas might start to eat healthier and more nutritious food (Roggema 2014a; Roggema et al. 2014).

7.3 Initial Stage

In August 2013 a family holiday brought us to Rio de Janeiro. Amaury Pugliese showed us around the city and brought us to Cantagalo. It was a dark evening as we climbed the stairs and reached the favela. We had a great view (Fig. 7.1).

We visited the jiu-jitsu school of Team VB. High up in the favela about 40 youngsters are training every day for 1.5 h (Fig. 7.2). They are taught the tricks of

Fig. 7.1 Cantagalo view (*photo* Rob Roggema)

Fig. 7.2 Young kids in the jiu-jitsu school (*photo* Rob Roggema)

the trade of jiu-jitsu, but also elementary behaviour such as being on time, have a regular life, follow instructions and formulate an ambition in life. The training session is hard work, the discipline is high and the kids return home exhausted. Looking at the intensity, the sweat plunged and the focus in the eyes of the kids and one wonders how on earth will they get enough food to compensate all the Joules burnt. Here the initial idea emerged. The answer on my question how we could help these kids was that if we could send a bag of beans and a bag of rice every month, they would have enough to eat. But, as an old proverb says: 'instead of bringing the people fish you'd better learn them to fish'. Analogue to this, the idea was born not to bring them bags of rice or beans, but to teach the people how to grow their own food. This way food can be produced close to where the demand is. This gave direction to the development of the first ideas how to implement productive spaces in the favela.

7.4 Analysis and Conceptualising the Idea

Upon our return in the Netherlands the idea began to ripe. What if we could build something where the people in the favela could grow their own food? The first task was to embed the idea in the University. After talks with the Board and the management, the idea could be further explored and a fact-finding mission was supported. This support was important to further develop the idea. On the basis of the

Fig. 7.3 Presentation WECANTAGALO

local experience and Google-maps, a spatial analysis and conceptual design for the entire favela could be developed. The elevation, the water system, spatial structure and density of built-up area, flat roofs and public spaces were determined and a systemic design proposition for locating water basins, FoodRoofs and potentials for solar power was created (for the maps see Chap. 6). This story of an approach, in the form of a presentation called WECANTAGALO (Fig. 7.3), formed the basis for many conversations in Rio de Janeiro.

7.5 Follow up Visits

With the conceptualised design proposition in mind several visits to Rio de Janeiro have been undertaken. During the first visit several meetings and new contacts were established.

First of all the favela was visited. Together with Amaury and Rafael (Fig. 7.4), a resident of the favela, the area was inspected.

We climbed the hill overlooking the favela (Fig. 7.5) and saw the houses were well constructed (Fig. 7.7 and 7.8), ready for an additional floors (Fig. 7.15). This gave us the confidence that putting heavy water-tanks on the roofs wouldn't be a problem. Moreover, many roofs were flat (Fig. 7.6, 7.8 and 7.13), normally in use

Fig. 7.4 Amaury and Rafael (*photo* Rob Roggema)

Fig. 7.5 View over Cantagalo (*photo* Rob Roggema)

Fig. 7.6 Many flat roofs (*photo* Rob Roggema)

Fig. 7.7 Strong construction, ready for a next storey (*photo* Rob Roggema)

Fig. 7.8 More strong and flat roofs (*photo* Rob Roggema)

for drying the wash or the regular barbecue. Some of the roofs have the best views of Rio de Janeiro (Fig. 7.14).

The houses are built steep up the hill, making rising stairways necessary (Fig. 7.9). On the roofs the current water provision tanks are stored (Fig. 7.10). In the tanks clean (chlorified) water is stored for use in the kitchen and daily household use.

Alongside the steep alleys ingenious gutters are fabricated (Figs. 7.11 and 7.12), meant to rapidly discharge the rainwater. Water is discharged under and in between the houses, taking the garbage with it downhill (Figs. 7.13, 7.14, 7.15 and 7.16).

Occasionally, a shop appears in the middle of the crossroads of alleys, streets and dense housing. These shops (Fig. 7.17), sell food, but only pre-packaged candy, potato chips and cakes, and lemonade and water. Healthy food is absent.

One of the potential roofs suitable for a FoodRoof was then top of the house of Rafael Lezinho (Fig. 7.18). We visited his place several times. The first time we measured the space available. An odd-shaped space, which was in use for washing, some plants, water tanks and the barbecue (Fig. 7.19, 7.20, 7.21).

After we created the first design propositions we returned to the house to discuss the plans (Fig. 7.22) and to explain the working of the aquaponic system and its elements (Fig. 7.23).

Crucial for the success of the FoodRoof is to gain political support. Therefore we had a meeting with the secretary of State for Sports and Leisure in the State of Rio de Janeiro, André Lazaroni (Fig. 7.23). We explained the benefits, presented the plans and asked him whether he could support a pilot project. He was very supportive of the idea and with the elections in the back his mind he agreed to finance the first FoodRoof. This was an important step in the process, because now we could present not only an idea and a purposeful and sustainable development, but also a perspective on realizing such a FoodRoof.

During our visits we also kept a close relationship with the Dutch consul in Rio de Janeiro. The consulate cannot support these kinds of projects financially, but the people working for the Consulate were very helpful in opening up their networks for us, so we could tap into the knowledge and support of others who were

Fig. 7.9 Steep streets and gutters facilitating rainwater discharge (*photo* Rob Roggema)

interested in FoodRoof-type of projects. We came in contact with some architectural firms, such as RUA-lab with Pedro Evora, and people in the academic scene.

With colleagues from the University we visited Cantagalo once again and discovered more suitable roofs (Fig. 7.24), the Cantagalo elevator, which brings people and garbage up and down the hill (Fig. 7.25) and concluded that the closest

Fig. 7.10 Water tanks on the flat roofs (*photo* Rob Roggema)

food stall was located outside of the favela (Fig. 7.26). All in all, this brought us enough confidence to decide on progressing with the project. We had arranged political support, some finance, and many people were very positive about the idea. Also we found the urgency of providing the people living in Cantagalo with healthy food crystal-clear. Reasons enough to contact the EMOP, the state government agency, which manages the Acceleration program (PAC, Fig. 7.27) for cleaning up the favelas (see Chap. 4).

At the office of EMOP, we met Ruth Jurberg and her colleagues, and several representatives of Cantagalo residents. Luiz Carlos Dionísio (Associação de Ação Social Comunitária) and Luiz Bezerra do Nascimento (Associação de Moradores do Cantagalo) headed up the associations of local communities. We presented our ideas and both EMOP as the resident groups were very positive about the idea to create a series of FoodRoofs in Cantagalo. One leader of the community groups even asked us whether we could start the building of the roofs the same day. EMOP suggested to incorporate the FoodRoof requirements in the program of demands they are normally write for builders to bid for a certain work in the favela. This way the FoodRoof would become a normal feature that has to be realised by the builder who won the contract.

The last meeting we had was with the City government of Rio de Janeiro. The department of Public Works carries out a project called Hortas Cariocas, which aims at greening the cities gardens, and provides seeds for planting and crops. The government official explained all about this project, which at least consisted of an enormous row of project files (Fig. 7.28), but it remained unclear whether many

Fig. 7.11 Alley, also gutter (*photo* Rob Roggema)

Fig. 7.12 Stairway and gutter (*photo* Rob Roggema)

Fig. 7.13 Flat roof suitable for growing food (*photo* Rob Roggema)

Fig. 7.14 Roof with a view (*photo* Rob Roggema)

Fig. 7.15 Multiple storey housing (*photo* Rob Roggema)

Fig. 7.16 Discharging rainwater (*photo* Rob Roggema)

projects had been realized. A successful component of this project was the provision of seeds to the public. People can pick sacks of seeds up at the office, wrapped in pieces of old newspapers (Fig. 7.29). The people take these seeds home and plant them in their own garden.

Fig. 7.17 Typical food store (*photo* Rob Roggema)

Fig. 7.18 With Rafael Lezinho (*photo* Rob Roggema)

Fig. 7.19 Rafaels Roof (*photo* Rob Roggema)

Fig. 7.20 Rafaels Roof (*photo* Rob Roggema)

Fig. 7.21 Discussing the design (*photo* Bruno Fonseca)

Fig. 7.22 Explaining the manual (*photo* Bruno Fonseca)

Fig. 7.23 Together with André Lazaroni, Secretary of State for Sports and Leisure, Sate of Rio de Janeiro (*photo* Amaury Pugliese)

Fig. 7.24 Another suitable roof

Fig. 7.25 The elevator in Cantagalo

During these inventory visits the idea of a FoodRoof was explained to as many interested organisations as possible, such as the Dutch Consulate, the State of Rio de Janeiro, EMOP, the city department of Public Works. Residents were contacted and the possibilities were discussed. The idea could count on wide support and this gave reason to intensify the design process. Two graduates of the University were approached to elaborate the design further.

Fig. 7.26 Local vegetable shop outside the favela

Fig. 7.27 Construction PAC billboard

Fig. 7.28 The Hortas Cariocas project files (*photo* Rob Roggema)

Fig. 7.29 The seeds pre-packed in small papers to be given out to residents (*photo* Rob Roggema)

7.6 Design Process

The graduates started to design the FoodRoof further and detailed the design technically and materialised the proposition (see Chap. 8). This phase was conducted mostly in the Netherlands, with frequent exchanges by email with the partners in Rio and the house owners in Cantagalo favela.

Introducing a food producing system in the favela comes with certain specific requirements. To be acceptable for the residents it has to be a system, which must be productive, is lightweight and easy to dismantle, closes cycles and is sustainable, and therefore makes use of proven technology. For these reasons an aquaponic system was chosen, which produces food through a sequence of fish tanks, worms, vegetable hydroponics and a vertical gutter system. The fish enrich the water with nutrients, which are filtered and modified by the worms in order to be taken up by the roots of the plants. This water based system uses the water as a carrier for nutrients after which the plants extract the nutrients from the water hence purifying the water before it is pumped back into the fish-tanks. This way the cycles of water, nutrients and waste are closed and with help of solar power the energy cycle too. There are already several good examples of realised systems, such as the biospheric project in Manchester (Keeffe 2014; Roggema 2014b), and projects in Amsterdam and Rotterdam in the Netherlands (Broekaert 2012). In order to let the system function, water is required. In Rio, the annual rainfall is more than sufficient, but in order to use the water it needs to be captured and locally stored, preferably on the roof where it will be used.

Fig. 7.30 FoodRoof manual

This system is not very complex, but for ordinary citizens also not very simple to understand. In order to help residents to construct a FoodRoof by themselves, we have developed a manual (Drissen and Broekhuis 2014), giving explicit, visual and clear directions on how to construct the roof step-by-step. This bilingual (English and Portugese) manual (Fig. 7.30) contains easy to understand 'IKEA-like' information with images of tools and materials needed for the construction.

The design was finalised with construction drawings, illustrating the dimensions and materials required for building the FoodRoof in detail (see Chap. 8). Meanwhile, local support, financially and mentally, was sought to implement the project.

7.7 Building the First Roof

The next phase consisted of the realisation of the project in Cantagalo. A team of builders, amongst which the two graduates, the local architect from EMOP, and the author, spend a week in the favela to build the FoodRoof. After this the operation of the roof is evaluated and steps for scaling up and follow up are taken. This first FoodRoof was built in 2014. It turned out that the roof of Rafael was not available at the time of building, but an alternative was quickly found. The roof of Marcelo Açunçao is located in the heart of the favela, alongside a busy street. In five days the materials have been bought and transported to the roof, where the fish-tanks are painted in catchy colours, the vertical gutters were put together and the horizontal vegetable plant-bed was installed. The tubes, pipes and pumps are connected to start the system. In the fish-tanks a total of 40 tilapias live, strawberry, pepperoni and tomatoes are grown in the plant-bed, and a range of herbs are planted in the vertical system. After five days the system was complete. During the building many residents from the neighbourhood came over on the roof to help with the construction and helped out carrying stuff. By the end of the week the system was operational. After a week or two the system should be in balance and could be maintained with simple actions.

7.7.1 Evaluation

The evaluation of the first realised FoodRoof shows both successes and failures. The FoodRoof produces food and the system works. It prevents dirty water from ending up in Guanabara Bay and the house-owner and residents are enthusiastic. The main problem after realising the FoodRoof was the summer heat in Rio. The plants and fish had difficulties to deal with temperatures above 45 degrees and scorching solar radiation. The proposed sun cover of solar panels couldn't be realised, but would still be a useful solution for the heat and it would close the energy cycle, as the pumps currently retrieve their electricity from the grid.

The materials were obtained from the local shop and materials shop. Besides being expensive the use of new materials do not close the material cycle, and it would be better if the FoodRoof could be build with waste materials. Finally, the local embeddednes could be stronger. After the FoodRoof was build the positive vibe amongst the residents could have been used to start up more FoodRoofs in Cantagalo.

The research project is action research based and practice-oriented, and took place in an emergent and dynamic process. The question regarding the research approach is whether it should be structured beforehand, or if emergent ideas should inform the approach during the research. This action research project has profited from emerging ideas, it has brought new insights, new connections and new opportunities to realise the roof. This mutual information informed both the research project and practice. It proofed to be very fruitful, and without it the first FoodRoof would have probably never been build.

Time will learn whether the roof will be used in the long term. For now, the owner is very happy with the roof and enthusiastic about the system. When this first FoodRoof remains the only one, it is deemed to end as a lone wolve: a strong proposition with no followers. The FoodRoof requires a follow up at larger scale, in order to profit from exchange of learning and the economies of scale in the production, while still being a solution for individual houses.

It can be concluded that it is possible to develop a roof that provides the food for the family that lives underneath it. Local residents support the idea, but lack the organisation to extend the project, they seem to be a bit apathetic and refuse to take initiative if it doesn't bring them direct profits. A rigorous research plan without risks would not have been successful in developing the FoodRoof. It is necessary to take a risk in the process in order to make things happen. Imagine, the research had been eloquent and the researcher had been waiting for acceptance of their ideas, the first FoodRoof still had to be build.

References

Barchfield, J. (2013). *Brazil's favela foodies go off beaten path in Rio's slums*. Published online: May 28, 2013. http://www.huffingtonpost.com/2013/05/27/brazil-favela-foodies-rio-slums_n_3343353.html?ir=India&adsSiteOverride=in

Broekaert, J. (2012). Planten kweken op vissenpoep. *NRC Next*, September 6, 2012. Published online: http://www.nrcnext.nl/koken/2012/09/06/planten-kweken-op-vissenpoep/

BUILD (Bombay Urban Industrial League for Development). (2010). Eat less, sleep less, and work more: Situational analysis of food insecurity in urban slums of Mumbai. *Bread for the World*, Germany. http://buildindia.org/FINAL%20REPORT.pdf

Consulaat Generaal van het Koninkrijk der Nederlanden. (2014). *Design na Praça - Ontwerpen op het plein. Denken, doen en debatteren over design met afval. Nieuwsbericht*. Published online: November 6, 2014. http://riodejaneiro.nlconsulaat.org/nieuws/2014/11/design-na-praca.html

Drissen, M., & Broekhuis, B. (2014). *Food growing roof terrace; Instruction manual aquaponic system*. Velp: VHL University of Applied Sciences.

Faye, O., Baschieri, A., Falkingham, J., & Muindi, K. (2011). Hunger and food insecurity in Nairobi's slums: An assessment using IRT models. *Journal of Urban Health, 88*(Suppl 2), 235–255. doi:10.1007/s11524-010-9521-x

Flavors of Brazil. (2012). *Slumming it, dining in a favela*. Published online: January 10, 2012. http://flavorsofbrazil.blogspot.in/2012/01/slumming-it-dining-in-favela.html

Keeffe, G. (2014). Urban agriculture: Productive synergies in the city. In: R. Roggema & G. Keeffe (Eds.), *Why we need small cows. Ways to design for urban agriculture* (pp. 75–98). Velp: VHL Press.

Kimani-Murage, E. W., Schofield, L., Wekesah, F., Mohamed, S., Mberu, B., Ettarh, R., et al. (2014). Vulnerability to food insecurity in urban slums: experiences from Nairobi. *Kenya. Journal of Urban Health, 91*(6), 1098–1113. doi:10.1007/s11524-014-9894-3

Mohan, S. (2015). *Junk food is feeding a malnutrition epidemic in Delhi's slums*. Published online: June 1, 2015: http://scroll.in/article/728525/junk-food-is-feeding-a-malnutrition-epidemic-in-delhis-slums

Petrini, C. (2014). The favela chef turning food waste into organic dishes. Published online: February 11, 2014. http://www.slowfood.com/international/food-for-thought/focus/209868/the-favela-chef-turning-food-waste-into-organic-dishes/q=E839E5

Roggema, R. (2014a). FoodRoof Rio: How favela residents grow their own food. *Adjacent Government, 4*(2014), 18–21.

Roggema, R. (Ed.). (2014b). *Greg Keeffe: The nutritious city*. The Biospheric Foundation. International Urban Agriculture Lecture, March 5, 2014. Velp: VHL University of Applied Sciences. ISBN: 978-90-822451-0-3. 51 pp.

Roggema, R., Pugliese, A., Drissen, M., & Broekhuis, B. (2014). The FoodRoof: How Cantagalo and Pavão-Pavãozinho Favelas Grow Their Own Food. In: R. Roggema & G. Keeffe (Eds.) *Why we need small cows. Ways to design for urban agriculture* (pp. 207–230). Velp: VHL Press.

Rothman, L. (2014). *Rio's favelas are feeding Brazil's elite*. Published online: November 3, 2014. http://munchies.vice.com/articles/rios-favelas-are-feeding-brazils-elite

Chapter 8
The Design of the FoodRoof

Rob Roggema

8.1 Introduction

The FoodRoof project is a private development initiative, which started in March 2014. A first demonstration roof was realised in September 2014. The idea for the FoodRoof project emerged during a visit to the favela (slum area) called Cantagalo in Rio de Janeiro. Food consumed here is often not very healthy and the intake of fresh vegetables and quality meat or fish is minimal. What if we could teach the people in the favela to grow healthy food in their own environment? A design project was started and two graduates elaborated the design for the FoodRoof, made the construction drawings and materialisation details, and developed a construction manual. As we have seen in chapter seven, the FoodRoof idea has been presented to the governor of the State of Rio de Janeiro, the Provincial Programa Aceleração do Crescimento (PAC, the Accelerating Growth Programme), which takes care of construction works providing clean water, waste management, road infrastructure and sewage in the favela's, the Dutch Consul in Rio de Janeiro and two community groups from Cantagalo. All supported the FoodRoof idea. The first FoodRoof is only one single roof in a sea of flat roofs in Cantagalo. The purpose of this first roof is to prove that the system can function on a roof in a slum area in order to stimulate replication of this model to other roofs in this and other favela's.

R. Roggema (✉)
Faculty of Design, Architecture and Building, University of Technology Sydney, Ultimo, NSW, Australia
e-mail: rob@cittaideale.eu

R. Roggema (ed.), *Food Roofs of Rio de Janeiro*,
DOI 10.1007/978-3-319-56739-6_8

8.2 The FoodRoof

This FoodRoof consists of an aquaponic system, which is well known in many Western countries. The application of this technique in a new context, in challenging spatial and social conditions and offering access to healthy food for local residents, is truly innovative.

The FoodRoof is located in Cantagalo favela. The house of Marcelo Assunção is located half way the hill, in the middle of the favela. The location is embedded in the urban water system, discharging rainwater from the top of the hill downwards. Additionally, the rainwater that is falling on the roof of this particular house can be used to supply water to the food producing system. The house is also served with clean drinking water, which is provided from the city water company, and is stored on the roofs in 1000 litre water tanks, 3–4 on each roof. In order to create a food-growing roof in this dense part of the favela all the space on roof needs to be used for the growth of food.

The principles used for the design of the FoodRoof aim for a sustainable use, now and in the future. The cycles of water and nutrients are closed, waste is used as a resource and energy is generated with a small solar panel. The plants are placed in the scheme where they will get enough sunlight, but stay out of the strong direct sun during summer. This system is inspired by the Biospheric Foundation project, developed for the Manchester International Festival 2013 (http://www.biosphericfoundation.com; Roggema 2014), by Vinny Walsh and Greg Keeffe. The core characteristics have been used to elaborate the early sketches and transform them into a detailed constructive design (Broekhuis and Drissen 2014). The proposed system is called Biospheric Aquaponic System (Fig. 8.1). This system consists of several fish-tanks, two grow-beds for plants and a solar pump to circulate the water. The plants grow in hydro-grains and are supplied with (rain)water.

The fish-tanks contain fish, which produce waste in the form of ammonia.

Water with ammonia is led through pipes to the grow-bed for plants. Plants require water, light, CO_2 and bacteria like nitrates. The bacteria in the grow-bed break the ammonia down to nitrates, which the plants subsequently use. The plants extract nitrates from the water and filtrate the water, when it is passing through the grow-beds. The solar pump is needed to circulate the water through the entire system and return water to the fish-tanks. Hence this creates a closed cycle. Residents eat the vegetables and fish (as soon as the fish is big enough and ready for consumption). Their food waste feed the fish. This fish food must contain a lot of proteins, which are found in the waste from food such as bread, grain, fish waste, maggots or worms. This means there is no additional input needed from outside the system, besides purchase of materials, plants and small fish, and the capacity to realize the FoodRoof. Every month the system loses 10–15% of water due to evaporation. When water is added to the system it shouldn't be taken from chlorinated city water, but instead, the rainwater is of sufficient quality and could complement the water-tanks. The power required for the pump will be supplied using a small solar panel.

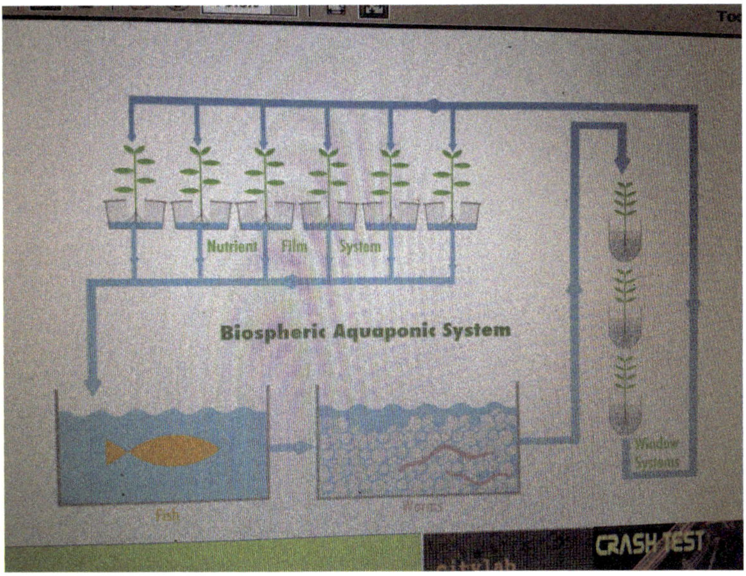

Fig. 8.1 Biospheric Aquaponic System. *Source* Greg Keeffe

8.3 Early Sketches

The early sketches illustrate the size of the roof terrace, approximately 8 × 5.5 meters and how it is currently used: to store water tanks, for laundry and to barbecue. It is an open space, with a great view over the rest of the favela and Ipanema and the ocean (Fig. 8.2).

Initially, two main options were available to transform the house into a FoodRoof (Fig. 8.3). In one option all the functionalities of the system were foreseen on the terrace of the house and in the other option the roof is used for water storage while the food is grown on the terrace.

With the water captured on the roof and transported to the fish-tank (Fig. 8.4), the water could be led further down into the vertical system, where it is led through pockets of substrate with plants (see reference image Fig. 8.5), before it enters the grow-bed and is pumped back upwards in the fish-tank (Fig. 8.6).

This design is a little adjusted (Fig. 8.5) with the water storage, fish-tank and grow-bed connected to each other on the roof. This design, which was originally made for the roof of Rafaels house, could easily be transformed to fit on the roof of Marcelo. Due to practical reasons Rafaels roof could not be used on time, so the first roof was realised on an alternative roof close by.

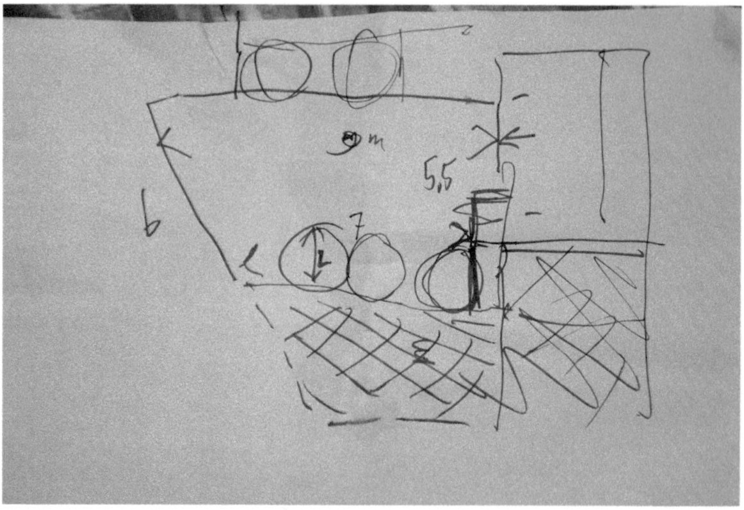

Fig. 8.2 First analysis

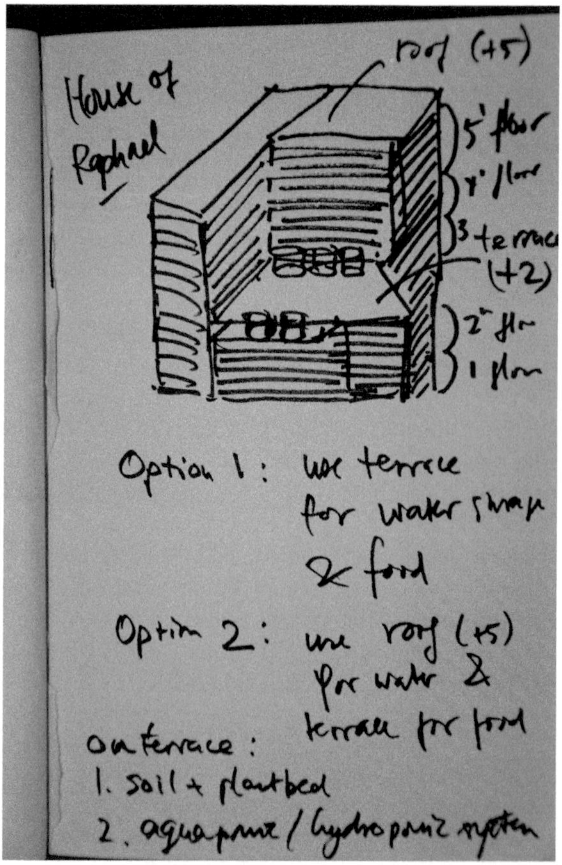

Fig. 8.3 Two design options

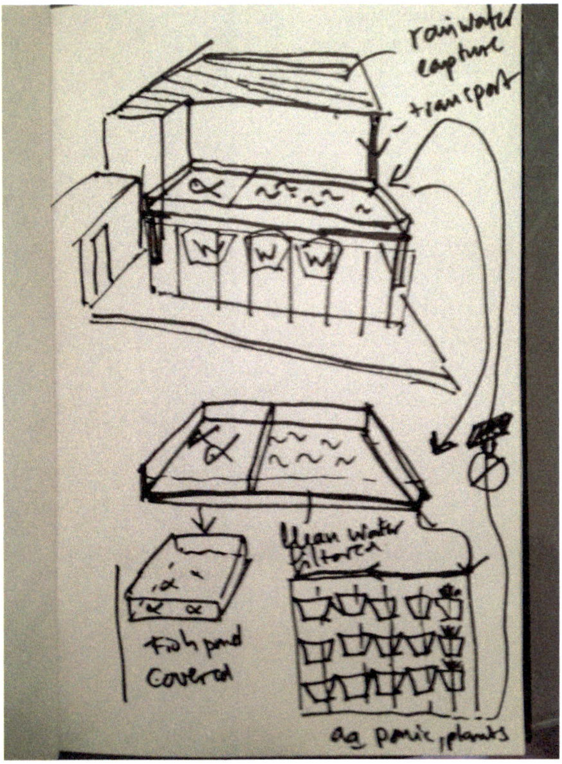

Fig. 8.4 Design sketch

8.4 The Design

The size of the roof is approximately 17.5 m^2. On the roof a simple aquaponic system is designed, consisting of three water tanks of 700 litres each with fish in different stages of maturity, two growing beds for plants (one horizontal growth bed of 3.5 m^2, one vertical plant system and a solar pump that constantly circulates the water (Fig. 8.7). The plants grow in hydro-grains and are supplied with water from the fish tanks (containing ammonia) that is led through pipes to the grow-beds where bacteria break the ammonia down to nitrates and make it available for the plants. After the plants have taken up the nitrates, the water is circulated back into the fish tanks. The sizes of the growth beds and volumes of fish tanks have been based on the estimated amounts of vegetables, herbs and fish that a family of five consumes.

The fish feed is obtained from protein rich food waste (such as bread, grain, fish residue, maggots or worms) being mainly residues from their own kitchen, which the family can collect while operating the FoodRoof, Where additional fresh water is needed for the fish tanks or for irrigation of the plants, fresh rainwater is used that is collected and stored in a collective water-tank located higher up in the favela (to

Fig. 8.5 Reference of vertical plantation. *Source* Greg Keeffe

be used by multiple FoodRoofs), also preventing that storm water runs through the favela collecting dirt and garbage on the way, ending up in Guanabara bay, where it pollutes the ocean.

The energy supply for the pump was designed to come from solar panels placed as a roof above the fish tanks and the growing beds to also function as a sun cover, but these have not been installed yet. The pumps currently retrieve their electricity from the grid but as soon as the solar panels will be installed, only energy from these panels will be used.

The demonstration FoodRoof has been built on a house that is located along one of the main roads of the favela, the fish tanks are painted in striking colours and the vertical plant system forms a distinctive green wall on the roof along the roadside, making the FoodRoof well visible from everywhere in the favela. At the same time the design is efficient and fit to the small space available on the roof.

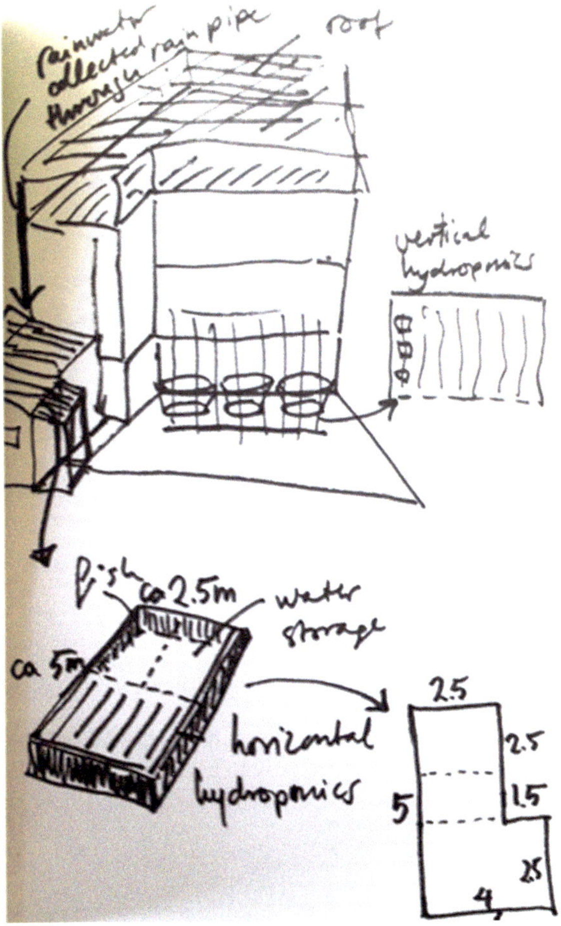

Fig. 8.6 Adjusted design

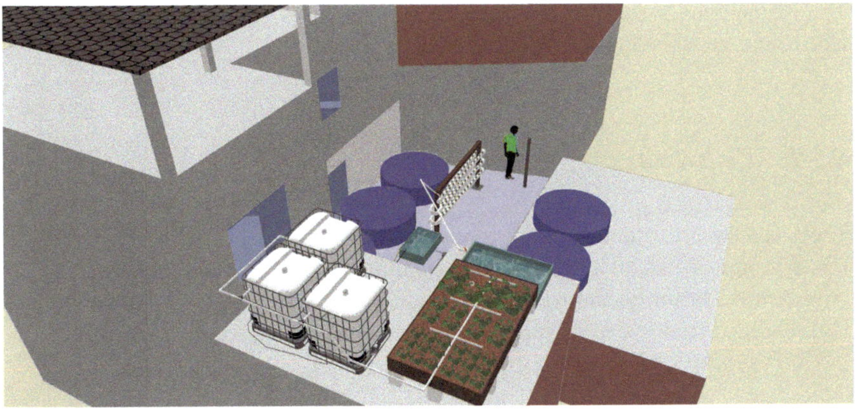

Fig. 8.7 Overview of the FoodRoof design

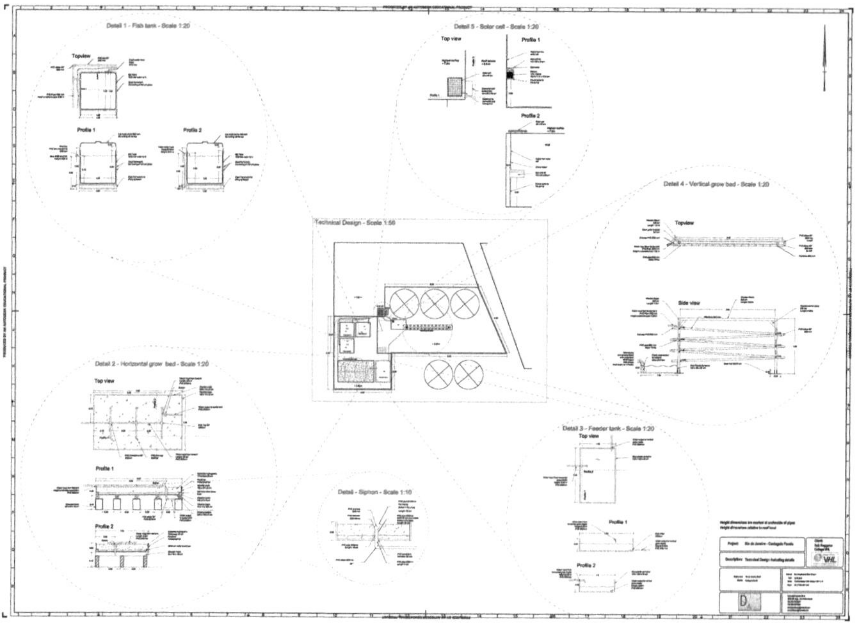

Fig. 8.8 Technical drawing of FoodRoof and details

Together with the technical design (Fig. 8.8), a manual was written explaining the principles of an aquaponic system, how to construct the FoodRoof, which tools and materials are needed, how to operate the system and how to solve frequent problems.

8.5 Components of the FoodRoof

The Aquaponic System on the roof consists of a several elements: Fish-tanks, a Horizontal grow-bed and a Vertical plant system.

8.5.1 Fish-Tanks

There are three fish tanks on the roof (Figs. 8.9, 8.10, 8.11 and 8.12). Every tank has a group of fish of different size: small/baby-fish, medium-growing fish, and large fish ready for consumption. Each tank has an overflow from where the water is discharged to one central pipe and is subsequently led to the grow-bed.

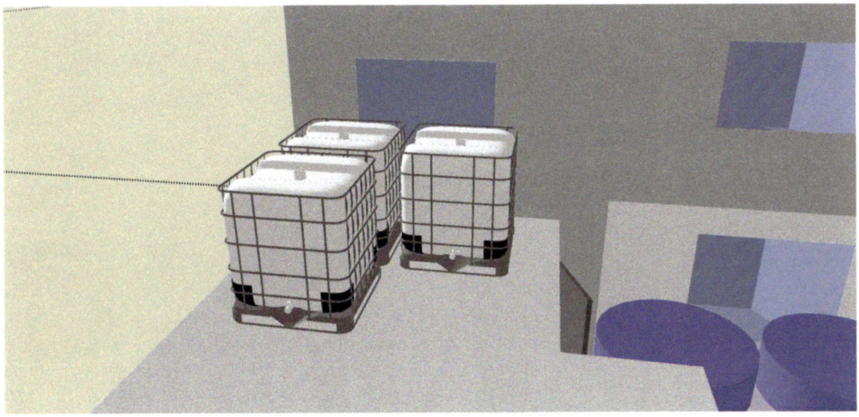

Fig. 8.9 Fishtanks

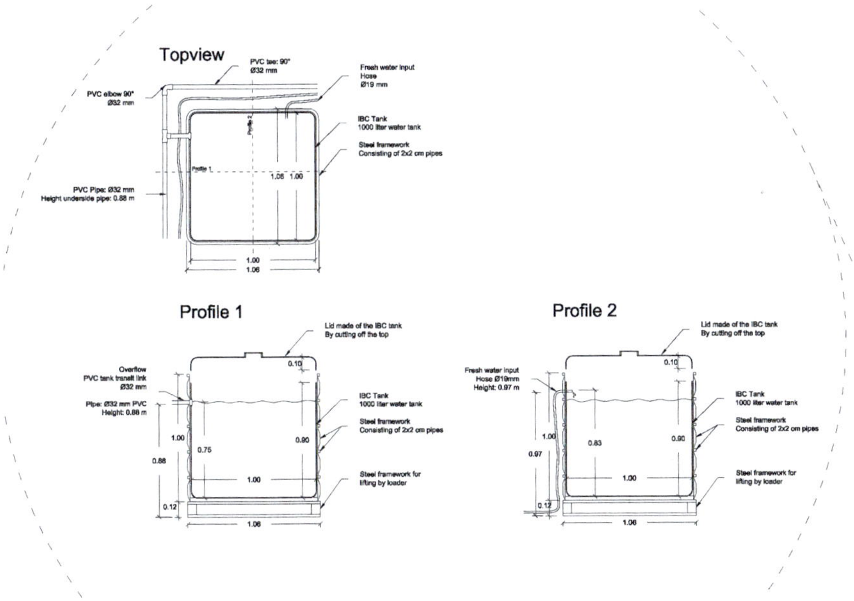

Fig. 8.10 Technical details fishtanks

The fish produces waste in the form of ammonia. Residents eat vegetables and fish (as soon as the fish is big enough and ready for consumption) from the system. The fish eat the food waste. The fish food contains a lot of proteins, which are found in the waste from food such as bread, grain, fish waste, maggots or worms.

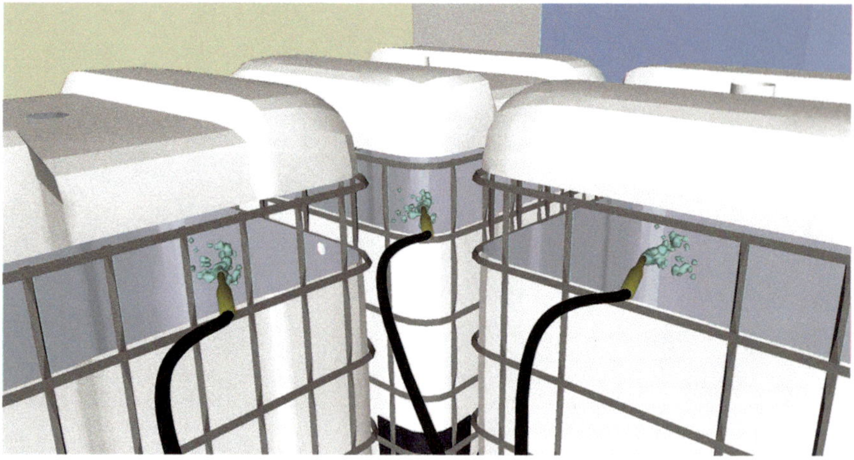

Fig. 8.11 Fill the tanks with water

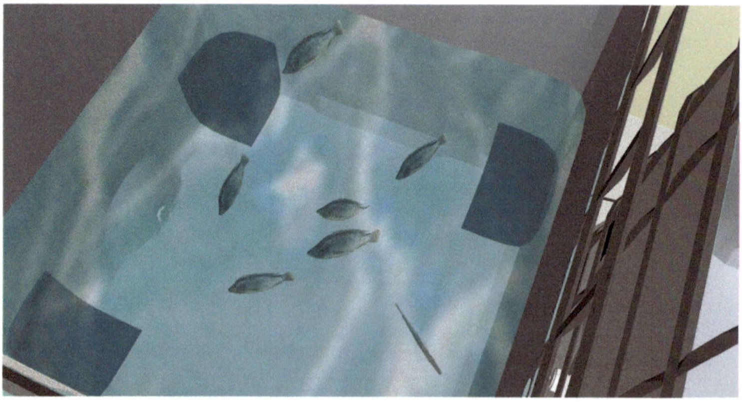

Fig. 8.12 Fill the tanks with fish

8.5.2 Horizontal Grow-Bed

The horizontal grow-bed (Figs. 8.13 and 8.14) is located next to the fish-tanks on the roof. This bed consists of a horizontal vegetable plant system. The water is irrigated through the bed in which the crops are planted on and in between clay balls. When the water reaches a certain maximum level it leaves the bed through a siphon (Figs. 8.15 and 8.16) to the feeder tank. The feeder tank is a storage tank for water from the horizontal grow-bed. Using a pipe on the bottom of the tank, the water flows continuously to the vertical grow-bed. From the feeder tank the water is pumped up to the upper tube in the vertical grow system.

Fig. 8.13 Horizontal grow-bed

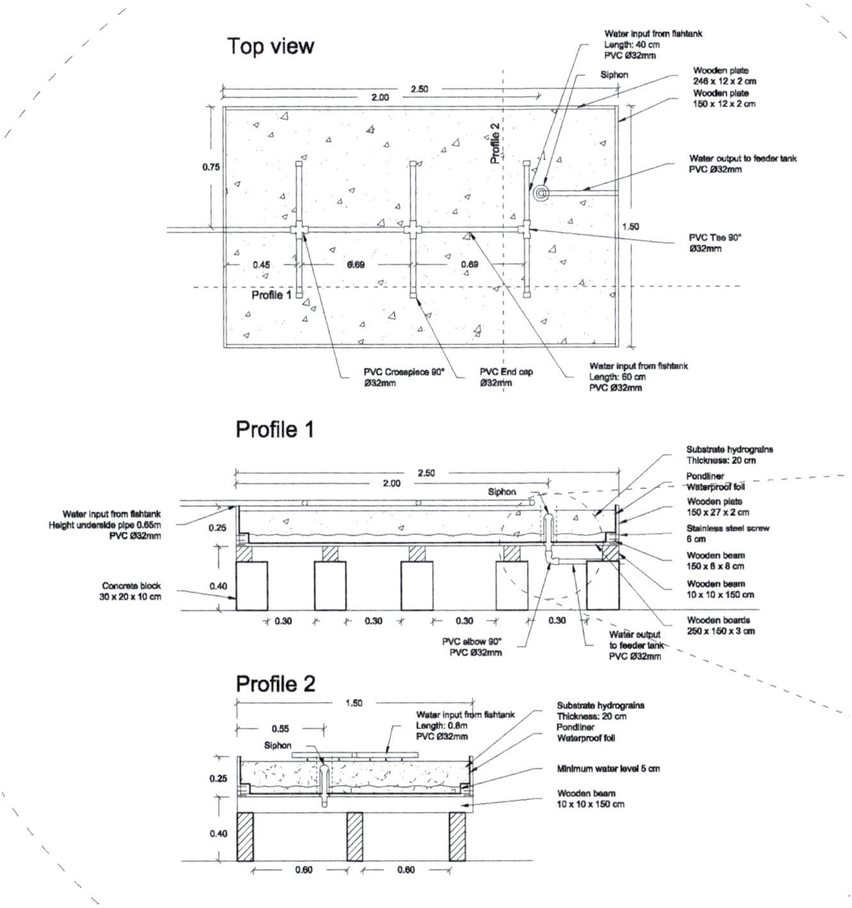

Fig. 8.14 Technical drawing horizontal grow-bed

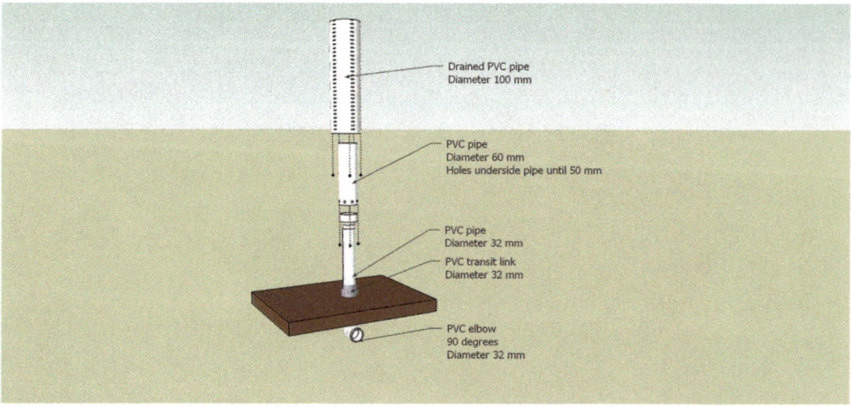

Fig. 8.15 Siphon

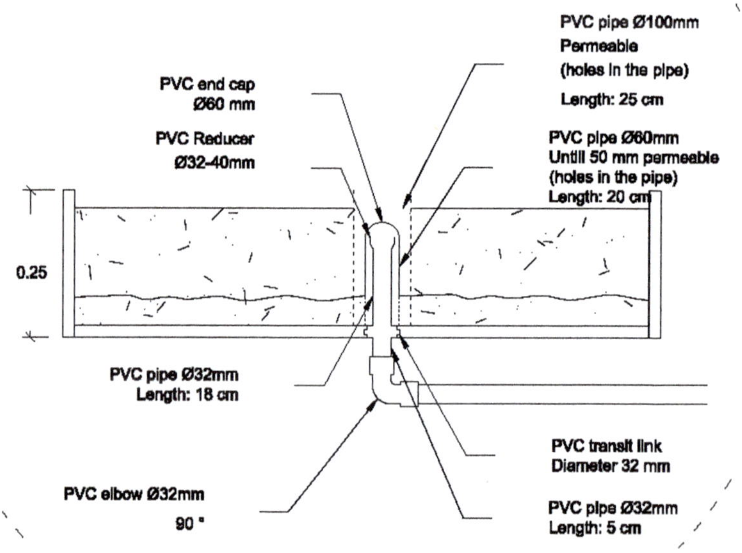

Fig. 8.16 Technical detail siphon

Water with ammonia is led through pipes to the grow bed for plants. Plants require water, light, CO_2 and nitrates. The bacteria which attach to the clay balls in the grow bed break the ammonia down to nitrates, which the plants subsequently use. The plants extract nitrates from the water and filtrate the water after passing the grow beds.

The crops that are proposed (Fig. 8.17) in the horizontal grow-bed are tomatoes, peppers, cucumbers, cabbage, eggplant, beans, bell pepper, zucchini, carrots and unions.

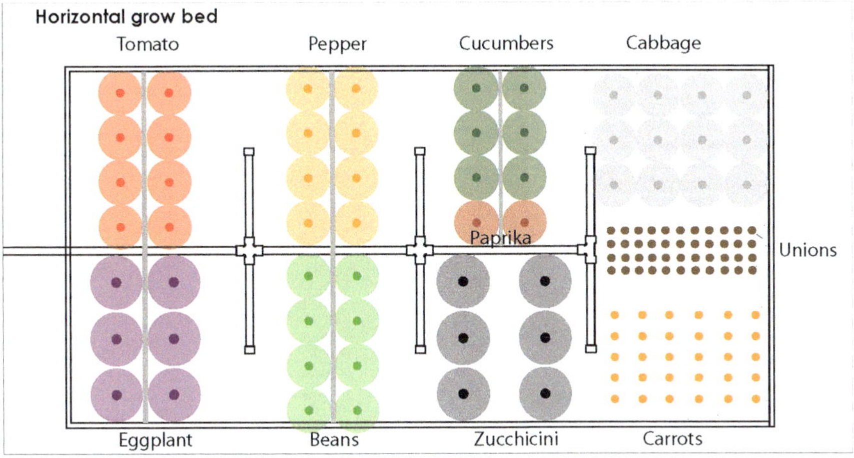

Fig. 8.17 Plant scheme horizontal grow-bed

8.5.3 *Vertical Grow System*

The second grow system (Figs. 8.18 and 8.19) consists of pipes through which water flows. The pipes have holes (see reference image, Fig. 8.20) in which the fruit and herb crops are planted in small substrate cups (Figs. 8.21 and 8.22). The roots of the plants must continuously be in the water. The water flows through the tubes into the pump-tank. A pump tank is placed at the end of the vertical grow system. A hose attached to the pump brings the water from the pump tank back to the fish tanks. The pump gets power from a 12 V battery, which is connected to a small solar panel. A solar pump is needed to circulate the water through the entire system and return water to the fish-tanks.

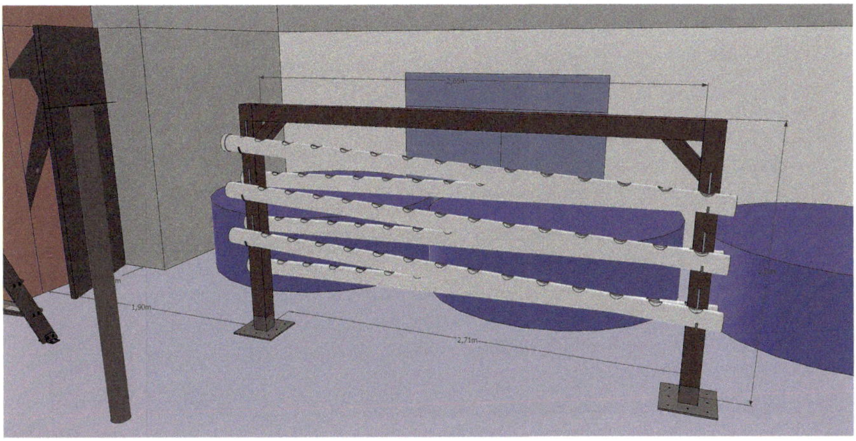

Fig. 8.18 Vertical grow-system

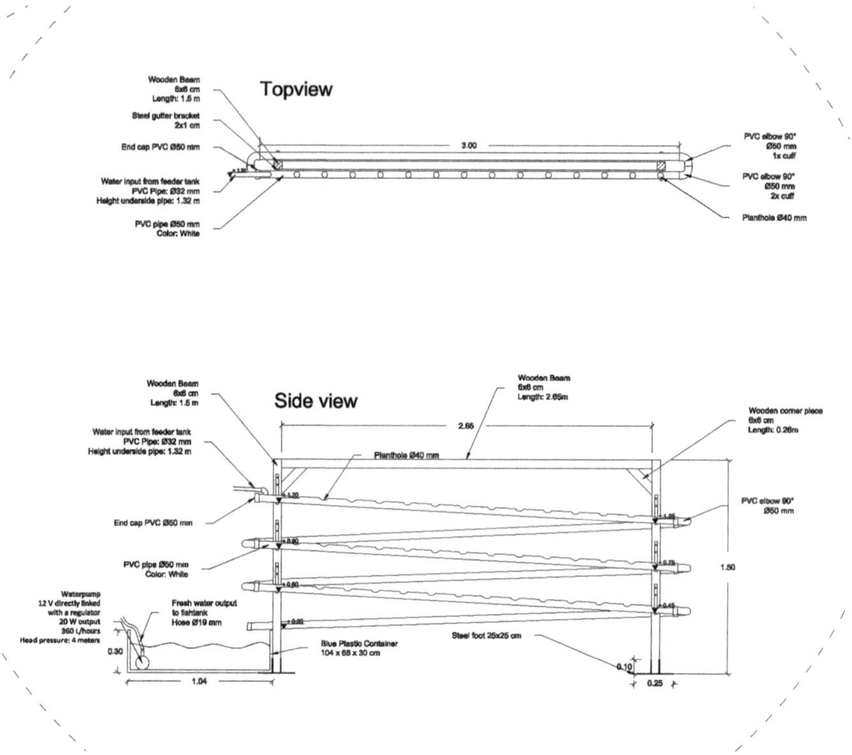

Fig. 8.19 Technical drawing vertical grow-system

Fig. 8.20 Reference of vertical grow-system

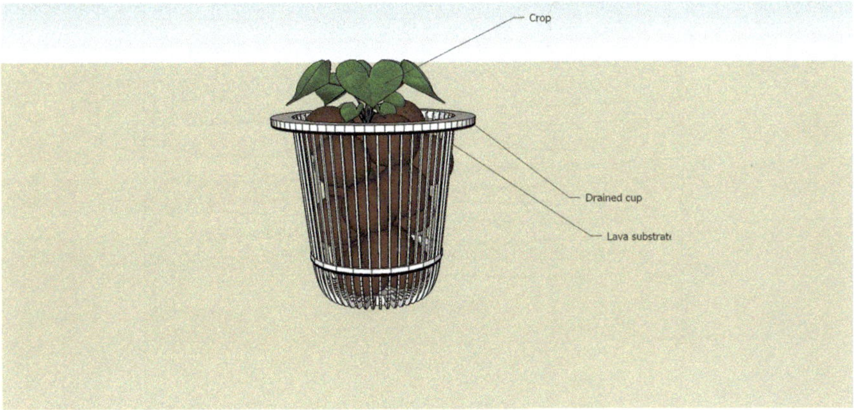

Fig. 8.21 Small cup with substrate and plant

Fig. 8.22 Cups fitted in the vertical system

The proposed crops for the vertical system are lightweight species, such as dandelion salad, lettuce, spinach, rocket salad, strawberries and chard (Fig. 8.23).

8.6 Test Installation

In the summer of 2014 a test installation has been developed to examine how the aquaponic system would work under warm climatic circumstances. The fish-tank and horizontal grow-bed have been super-positioned (Fig. 8.24) to make sure the

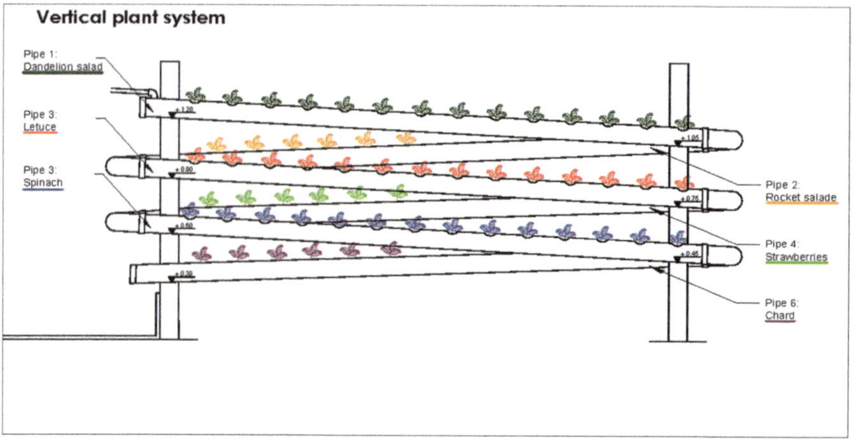

Fig. 8.23 Plant scheme vertical system

Fig. 8.24 Test installation

system works. The results were good. The system operated well and found its own natural balance within a couple of weeks. The system was easy to operate and could be adjusted through small changes by hand, whenever there was a problem.

8.7 Operation

The first FoodRoof was realised in one week. The initial start-up of the system went well, but several smaller adjustments were necessary.

The feeder tank (tank next to the horizontal bed where the siphon is connected to) and the pump tank (tank at the bottom of the vertical system) should be bigger. When the horizontal bed floes over, all the water goes to the feeder tank. This tank flows over and a bigger tank must be installed.

This feeder tank doesn't have to be very high, but it needs a large bottom surface to retain all the water. An important issue here is the pump needs to stay below the water all the time. If it sticks out of the water it will run dry. If that happens too often, the pump will stop working after a short period.

The feeder tank should be 1.20 × 0.50 m. The height should be at least 0.30 m. The pipe from the horizontal system entering the feeder tank should make the hole in the feeder tank waterproof.

The same dimensions are required for the pump tank. The PVC-pipe coming from the vertical system ends in the tank. The hole through which the PVC pipe fits should also be waterproof.

One of the fish-tanks is standing oblique. Place the fish tank on a horizontal surface. It should stand horizontal. This surface should be strong enough to carry the weight of the tanks.

Some of the PVC corner pieces are leaking. Use silicone kit or Teflon-tape to make these corner pieces waterproof.

When these adjustments are carried out the water circulation must be regulated:

The three fish tanks must be filled up to the edge with (rain)water. Then the valve in the pipe can be opened to fill up the horizontal grow bed. As soon as the bed reaches its maximum level, the siphon starts to work and empties the horizontal grow bed. The water flows into the feeder tank. The feeder tank doesn't overflow. While the feeder tank is being filled with water, the pump starts pumping the water into the vertical system. Regulate the pump in order to prevent it from running dry, through turning it closer or more open.

The pump tank at the end of the vertical system should also run when the vertical system is supplied with water. This pump should also be regulated so it never runs dry.

The regulation of the pumps is a process of trial and error and different methods should be explored. At the end of this exploration the pumps are regulated so the pump has never be turned off or on. The system can regulate itself. The easiest way to regulate the pumps is to link it to the speed of the siphon, which overflows every now and then. The pumps should be regulated according the interval between the overflows without the pumps running dry.

The system has to balance out within three weeks. After 3 weeks the system has created its own micro eco-climate. Bacteria are growing and the fish can be put in the fish tanks.

8.8 Lessons Learned

The evaluation of the first realised FoodRoof shows both successes and failures. The FoodRoof produces food, the system works and the house-owner and residents are enthusiastic. The main problem after realising the FoodRoof was the summer heat in Rio. The plants and fish had difficulties to deal with temperatures above 45 degrees and scorching solar radiation due to the fact that the proposed sun cover of solar panels couldn't be realised initially. It still is important to install the solar panels both to reduce the heat and to close the energy cycle. The solar panels can provide the energy for the household as well, but are expensive in Brazil, especially when compared to the illegally used electricity the favela houses generally use.

Several lessons can be learned from the development of this first FoodRoof:

- Work with the strengths within the community. The local community needs to be the owners of the project. When the local residents are made co-creators of the project they become committed, are open for collaboration and support and feel more responsible for maintenance; However, although local residents support the idea, they lack the organisation to extend the project, and they seem to be a bit apathetic and refuse to take initiative if it doesn't bring them direct profits.
- Use local food potentials and adapt the design to the roofs available, their size, their microclimate, natural shade, humidity, but also the place where rainwater can be captured and stored.
- Support establishment of a demonstration unit first and then—if successful—accelerate to extend the project beyond the initial pilot to the entire favela.

Since the first FoodRoof was build, several developments have added to the future prospectives of the FoodRoof:

1. A successful experiment has been undertaken—at the Design na Praça festival in November 2014—to build a FoodRoof system by using garbage retrieved from Guanabara Bay, demonstrating it is possible to create the system solely out of waste materials.
2. The uptake of the FoodRoof principle in the PAC programme of the State of Rio de Janeiro for retrofitting future favelas, which makes it an obligatory element in requirements for new building proposals.

3. The set-up of a new cooperative that aims to stimulate the realisation of FoodRoofs in Rio's favelas. The cooperative is a non-profit initiative of a group of Rio-based architects, social workers, and entrepreneurs together with Dutch and UK-based academics, bringing their networks together to support FoodRoofs in Rio.

8.9 Conclusions

8.9.1 Production

From the FoodRoof three types of produce can be harvested. The fish-tanks produce tilapia fish. In three stages the fish will grow from guppies to mature fish, which can be directly consumed. In the horizontal growth bed bell pepper, red pepper, strawberry and lettuce are cultivated. In the vertical plant system smaller plants, mainly herbs are grown. The family of five persons operating the FoodRoof is consuming all the produce. The maintenance of plants and fish and harvesting of products is carried out by the family members, a task that requires about 1–2 h a week. After the first year the system was restarted with seeds and fish stock bought by the operating family.

8.9.2 Economic Sustainability

The total costs of all materials for the FoodRoof, obtained in a local shop, was approximately €2500. The workforce required to build the FoodRoof (in one week) consisted of four people: a local architect, Marcelo Maurao, supported by the PAC program, two graduates and a professor of the University. The operating costs are minimal. Once in a while new puppies need to be bought and occasionally water needs to be added to the system.

To build the system with local workforce, the time needed for building is 96 h. With an average loan of 5.5 Real/hour, the total labour costs are 528 Reais, which equals approximately 135 euro. Solar panels need to be imported and will cost around 100 euro each.

8.9.3 The Societal Impacts

The FoodRoof provides the inhabitants of the house healthy and sufficient food. Compared to the diet most favela residents consume, which consists of candy, pre-wrapped cakes, lemonade and beer, the fresh vegetables and protein-rich tilapia mean a fundamental change in the diet of the family operating the FoodRoof, which they could otherwise not afford with their low income. The FoodRoof also reduces heat temperature in the building due to its insulating and reflecting effect.

Besides the nutritional improvements, the FoodRoof also had a positive impact on the community. Many residents come along to have a look and to asked questions, and the FoodRoof enhance local self-confidence and proud among the favela residents.

The application of aqauponic FoodRoofs, once it will be replicated on the many available flat roofs and a water harvesting system in the slum area, will also reduce the risks of flooding downstream and reduce contamination in the favela and in Guanabara Bay.

8.9.4 Policy Relevance

The FoodRoof is an accepted idea within the PAC program and it already became part of the program of demands for new construction projects in favelas. When this is realised the FoodRoof model has potentially a serious impact on improving the quality of life of favela residents. However, the current economic crisis has a negative impact on the realisation of more FoodRoofs in Cantagalo and other favelas in Rio de Janeiro and the political interest for this type of project has ebbed away. But once the economic and political crisis is over, the FoodRoofs could be successfully implemented as part of a broader urban policy in Rio de Janeiro and Federal policies in Brazil.

References

Broekhuis, B., & Drissen, M. (2014). *Food growing roof terrace; Instruction manual Aquaponic system, roof terrace Rafael Lezinho*. Velp: Van Hall Larenstein.

Roggema, R., Pugliese, A., Drissen, M., & Broekhuis, B. (2014). The FoodRoof: How Cantagalo and Pavão-Pavãozinho favelas grow their own food. In R. Roggema & G. Keeffe (Eds.), *Why we need small cows. Ways to design productive cities* (pp. 207–229).

Chapter 9
Building a FoodRoof in One Week

Rob Roggema and Marcelo Maurão

9.1 Introduction

The theoretical background and even the shopping list of required materials may be complete and ready this doesn't mean the construction of the FoodRoof is easy. We set ourselves a target of 'one-week, one-roof'. Within a week all materials needed to be purchased, tools needed to become available, the materials needed to be transported to the house and on to the roof and the system needed to be constructed. Without the help from local architect Marcelo Mourão, we wouldn't have been able to achieve the result. When the only available drill wasn't suitable, Marcelo brought his own, when the fish needed to be bought from outside the city Marcelo arranged the purchase and transport. When we were short on clay balls, Marcelo knew the flower stall where they provided us with 700 l. Meanwhile, two graduates

R. Roggema (✉)
Faculty of Design, Architecture and Building, University of Technology Sydney, Ultimo, NSW, Australia
e-mail: rob@cittaideale.eu

M. Maurão
Arquiteto Urbanista, Marqa Arquitetura, Rio de Janeiro, Brazil
e-mail: mmourao.arq@hotmail.com

R. Roggema (ed.), *Food Roofs of Rio de Janeiro*,
DOI 10.1007/978-3-319-56739-6_9

constructed the system on the roof. The manual proved to be an excellent help explaining to the installer, local residents and shop owners what was needed and how things should be constructed. When the system was nearly ready, there was no water available in the favela to fill the tanks.

Luckily the next morning the water supply was functioning again. On day 5 the system was in operation and ready to show to a delegation of the Dutch Consulate, the other favela residents, the State Government (EMOP) and colleagues and friends. The first FoodRoof in the world was realised.

The idea for the FoodRoof originates when Rob Roggema visited Cantagalo favela and discovered that the residents did not have access to healthy nor sufficient food. The idea was quickly born to support the people to grow their own food. By constructing a FoodRoof. From idea until realisation was however a big step. The preparation took six months. The design was delivered, calculations of the construction and the operation of the system were made. The amounts of fish, water, nutrients and the relations to each other were calculated and the required materials and tools were investigated. All the information was collected in a beautiful bilingual manual (Broekhuis and Drissen 2014). This manual formed the basis for the realisation of the FoodRoof.

What happened during the realisation? A story in five days.

9.2 Day One

On the first day we started with buying materials and tools. Together with Debora Bento, Marcelo Mourão, driver Jorge, all EMOP and Amaury Pugliese, we went to the huge building market Leroy Merlin (Fig. 9.3). We bought a range of materials (see box needed materials), such as pumps, many PVC pipes, tubes, a couple of clay balls, waterproofing paint, clips and rings and kit (Fig. 9.2). Besides these and other materials we bought some tools to work with, such as a saw, and pencils and markers (Fig. 9.1).

Needed tools

5

Tool: Saw

The saw is needed to saw the PVC pipes at the right dimensions.

The saw is also needed to saw the wooden plates at the right dimensions.

6

Tool: Hammer
The hammer is needed to hammer the nails into the wood.

7

Tool: Tapeline

The tapeline is needed to measure materials and distances.

8

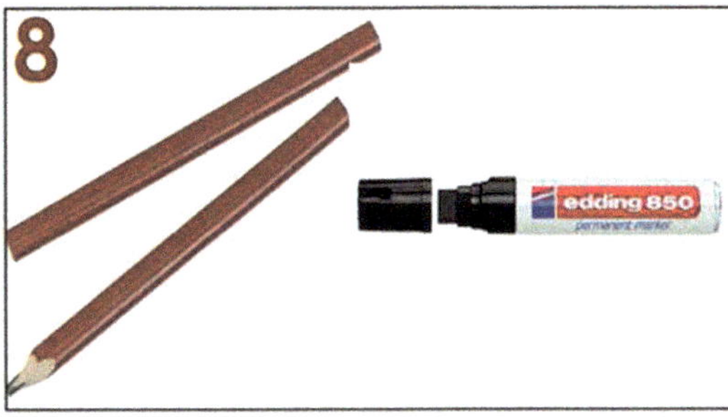

Tool: Pencil,
Permanent marker

The pencil and marker are needed to stripe on materials at the right dimensions.

Fig. 9.1 Tools required, page from the manual

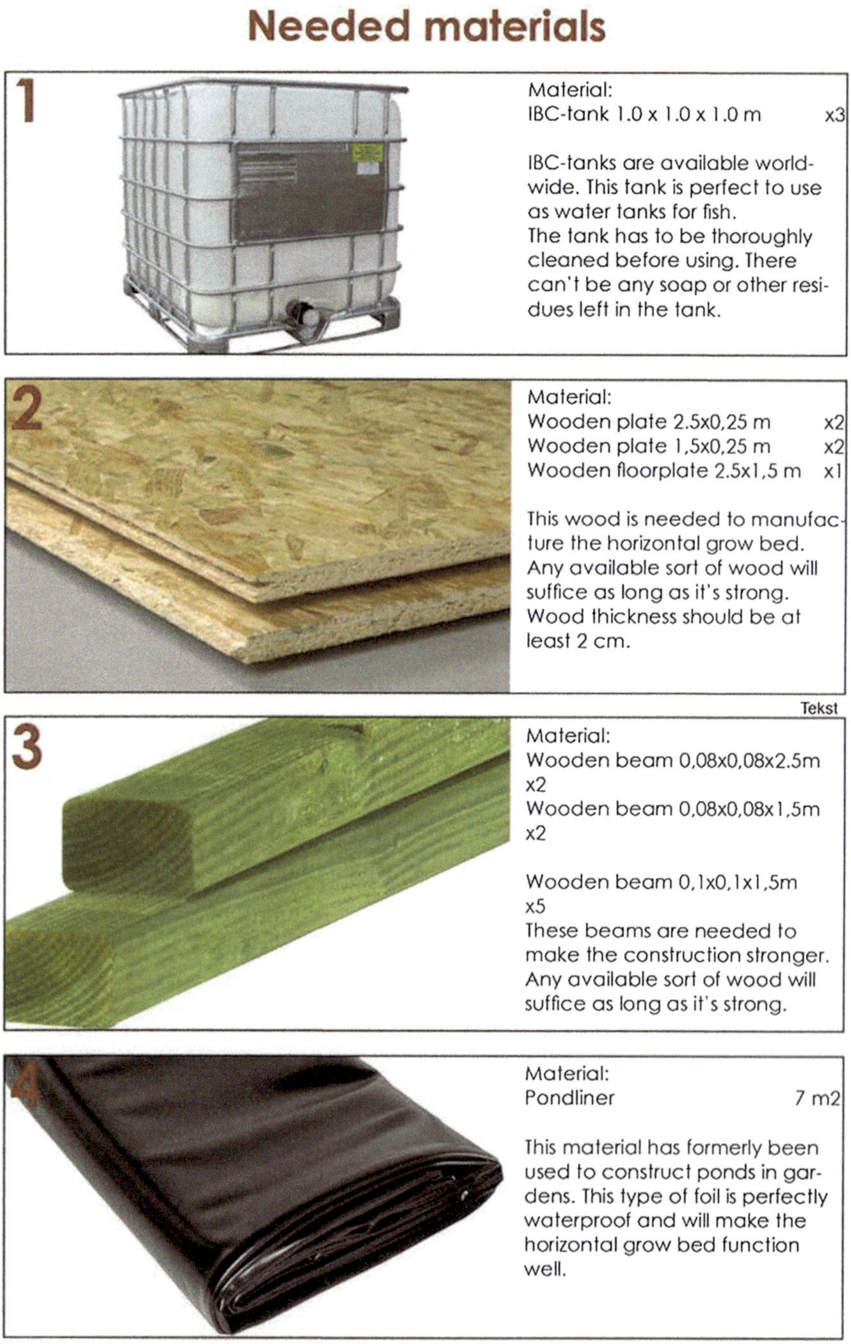

Needed materials

1

Material:
IBC-tank 1.0 x 1.0 x 1.0 m x3

IBC-tanks are available world-wide. This tank is perfect to use as water tanks for fish.
The tank has to be thoroughly cleaned before using. There can't be any soap or other residues left in the tank.

2

Material:
Wooden plate 2.5x0,25 m x2
Wooden plate 1,5x0,25 m x2
Wooden floorplate 2.5x1,5 m x1

This wood is needed to manufacture the horizontal grow bed. Any available sort of wood will suffice as long as it's strong. Wood thickness should be at least 2 cm.

Tekst

3

Material:
Wooden beam 0,08x0,08x2.5m x2
Wooden beam 0,08x0,08x1,5m x2

Wooden beam 0,1x0,1x1,5m x5
These beams are needed to make the construction stronger. Any available sort of wood will suffice as long as it's strong.

4

Material:
Pondliner 7 m2

This material has formerly been used to construct ponds in gardens. This type of foil is perfectly waterproof and will make the horizontal grow bed function well.

Fig. 9.2 Materials required, page from the manual

Fig. 9.3 Materials and tools shop (*photo* Rob Roggema)

Box—Needed Materials

Fish tanks		
IBC tanks		Quantity: 3
Horizontal grow beds		
Wooden floorplate	3.00 × 1.00 m	Quantity: 2
Wooden plate	3.00 × 0.25 m	Quantity: 4
Wooden plate	1.00 × 0.25 m	Quantity: 4
Wooden beam	0.08 × 0.08 × 3.00 m	Quantity: 4
Wooden beam	0.08 × 0.08 × 1.00 m	Quantity: 4
Pondliner m^2	3.5 × 1.5 m = 5.25 m^2	Quantity: 10,50
Concrete blocks	48 blocks under each bed	Quantity: 96
Hydrograins	0.6 m^3 each bed	Quantity: 1200 (L)
Feeder tank		
Plastic container	1.00 m × 1.00 m × 0.50 m	Quantity: 1
Solar cell and pump		
Solar cell		Quantity: 1
Battery	12 V, 25 AH, rechargeable	Quantity: 1
Battery box		Quantity: 1
Power cable meters	(meters)	Quantity: 10
Pump	20 W, 12 V, pump head: 2 m, Flow rate: 360 l/h	Quantity: 1
Garden hose meters	(meters)	Quantity: 14
Hose coupler		Quantity: 1
Piping		
PVC pipe meters	32 mm	Quantity: 20
PVC transit link	32 mm	Quantity: 3
PVC filter piece	32 mm	Quantity: 1
PVC elbow 90°	32 mm	Quantity: 5

(continued)

(continued)

PVC tee 90°	32 mm	Quantity: 5
PVC end cap	32 mm	Quantity: 12
PVC crosspiece	32 mm	Quantity: 4
PVC valve	32 mm	Quantity: 2
Siphon		
PVC pipe	100 mm, 0.3 m	Quantity: 2
PVC pipe	60 mm, 0.3 m	Quantity: 2
PVC endcap	60 mm	Quantity: 2
PVC pipe	32 mm, 0.4 m	Quantity: 2
PVC reducer	32–40 mm	Quantity: 2
PVC transit link	32 mm	Quantity: 2

We started this adventure in the building market chasing for PVC connections, tube fittings and many other pieces we need to build op the aquaponic system (Fig. 9.4).

Fig. 9.4 Looking for materials (*photo* Rob Roggema)

9.3 Day Two

On the second day our shopping hunger continued. This time we frequent some local building material shops in Cantagalo. These shops (Fig. 9.5), though way smaller than the huge building markets, offer even more materials. Additional PVC pipes, wooden poles of different sizes, huge Fortlev water-tanks, concrete blocks to support the horizontal grow-bed, and red and white paint all find their way to the house of Marcelo Assunção, who lives centrally in the neighbourhood. He has offered this roof for the experiment.

Marcelo Maurão, Deborah Bento Marc and Bart do the shopping (Fig. 9.6) and the graduates bring most of the materials to Marcelo Assunçã's house (Fig. 9.9). Marcelo Maurão orders the wood we need for the horizontal grow-bed, which they

Fig. 9.5 Local construction materials shop (*photo* Rob Roggema)

Fig. 9.6 Materials shop in Cantagalo (*photo* Rob Roggema)

Fig. 9.7 Inspection of the roof (*photo* Rob Roggema)

Fig. 9.8 Access to the roof

will deliver on Wednesday. In the afternoon we bring the materials bought yesterday and still in in the back of Amaury's car up the hill. Amaury leaves us quickly. He rather doesn't park his beautiful car in the neighbourhood.

On the roof we make an inventory and decide on the adjustments needed in the design for the realisation (Fig. 9.7) and how to bring all materials up the roof (Fig. 9.8). Afterwards we take a walk back down the hill (Fig. 9.9).

Fig. 9.9 Transporting materials to the roof

9.4 Day Three

On day three we finally made a start with the actual building. We start with the vertical system (Fig. 9.10). Drilling holes in the wooden poles, measuring the materials in detail and fitting the PVC pipes with the connecting pieces to each other (Fig. 9.11). The battery drill machine is not our biggest friend. It doesn't drill a single hole in the wood, so we do not make any progress. Moreover, we needed a screw bit but this wasn't available. After spending the whole afternoon searching for this piece we had to conclude that even with a drill bit the battery drill was to weak to drill decent holes or screw.

The wood, ordered on Tuesday for the horizontal grow-bed is still not delivered, so we cannot make a start building this. Because we need additional materials, we walk down to Ipanema to buy more support pieces and feeder tanks. We have success this time and we can buy everything we need. The only materials missing are the clay-balls, essential as substrate for the plants. We end up at a flower stall in

Fig. 9.10 Drilling the vertical system (*photo* Rob Roggema)

Ipanema (Fig. 9.12). It's always worth trying, isn't it? And to our surprise the owner takes us to the back of his little stall and there they are: big bags of the clay-balls we were searching for. We order 700 l (= 14 bags), which will be delivered on Thursday, just on time.

Upon our return at the house the three Fortlev water-tanks are just being delivered. They are huge, very blue and do not fit through the access hole to the roof. We need to tow them up on the roof (Figs. 9.13 and 9.14).

In the meantime the concrete blocks have been delivered too so we can outline the basis for the horizontal grow-bed (Fig. 9.15).

We decide to add red and white to the blue tanks, creating the Dutch flag on the future fish-tanks (Figs. 9.16, 9.17 and 9.18). The white paint is important to reflect the harsh summer sun. One by one the blue water-tanks are transformed in red-white-blue fish-tanks.

We paint the entire evening until the tanks are ready. In the evening light the red and white blink on the roof (Figs. 9.19 and 9.20).

A beautiful view when the night falls over Cantagalo (Fig. 9.21).

Fig. 9.11 Vertical system in its infancy (*photo* Rob Roggema)

Fig. 9.12 Flower stall in Ipanema, with clay balls (*photo* Rob Roggema)

Fig. 9.13 The watertanks have arrived (*photo* Rob Roggema)

Fig. 9.14 Watertanks towed up on the roof (*photo* Rob Roggema)

Fig. 9.15 The basis for the horizontal grow-bed (*photo* Rob Roggema)

Fig. 9.16 Water-tanks getting painted (*photo* Rob Roggema)

Fig. 9.17 Water-tanks in red, white and blue (*photo* Rob Roggema)

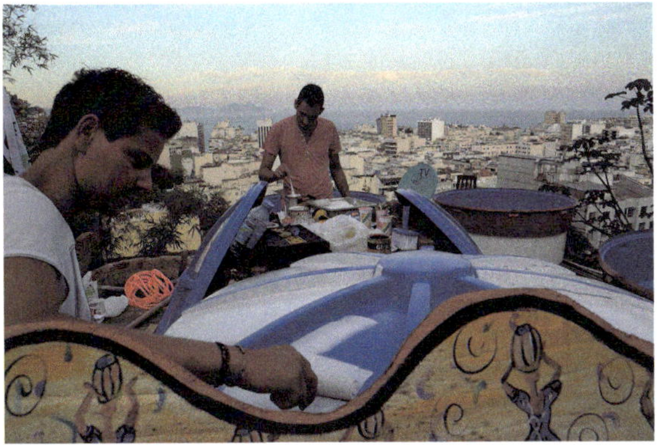

Fig. 9.18 Painting water-tanks (*photo* Rob Roggema)

Fig. 9.19 Painting the covers (*photo* Rob Roggema)

Fig. 9.20 Red, white and blue fish-tanks (*photo* Rob Roggema)

Fig. 9.21 The fish-tanks on the roof (*photo* Rob Roggema)

In the meantime several problems need to be solved. The borrowed drill still doesn't work. We can solve this because Marcelo Maurão brings his own, the next day. So, after another drill-less day we can finally start making holes on day Four.

9.5 Day Four

Upon arrival the next morning the FoodRoof starts to take shape already (Fig. 9.22).

Early morning the bags with clay-balls have been delivered and local residents help us to carry the bags through the tiny access up on the roof (Fig. 9.23).

Fig. 9.22 The roof, next morning (*photo* Rob Roggema)

Fig. 9.23 The clay-balls arriving (*photo* Rob Roggema)

But we received also bad news. The water supply to the favela is suspended and no one knows when it will be available. This is a problem, because the water is chlorinated and fish doesn't survive in it. We needed to have the water in the tanks for 24 h to let the chloride evaporate, but we won't have this amount of time left before the fish arrive tomorrow.

Marcelo Maurão just ordered the tilapia (Fig. 9.24), which needs to be brought in from 60 km out of town and will be delivered the next morning at 6 a.m.

Today Marcelo brought his own drill. This electric drill, for which we needed to buy an additional extension cord, functions well. The vertical system is constructed in no time. This system will be the vertical grow-system, in which smaller crops and herbs will grow (Figs. 9.25 and 9.26). The system in which, to our surprise the pipes hang crosswise in their clips (Fig. 9.27).

Finally, the long awaited wood for the horizontal grow-bed is delivered (Fig. 9.28), and is towed up to the roof (Fig. 9.29). Finally, we can start building this plant-bed.

The horizontal grow-bed is put together immediately. The side walls are sawed and even so the beams that support the corners of the plant-bed (Figs. 9.30 and 9.31).

Marcelo is painting the bottom of the horizontal plant-bed with waterproof paint (Fig. 9.32).

Marc perforates the siphon (Fig. 9.33) to make it ready for use in the horizontal plant-bed.

The piping system that brings the water from the fish-tanks to the horizontal plant-bed is finished (Figs. 9.34 and 9.35). Again, we work till late.

Fig. 9.24 Calling the tilapia supplier (*photo* Rob Roggema)

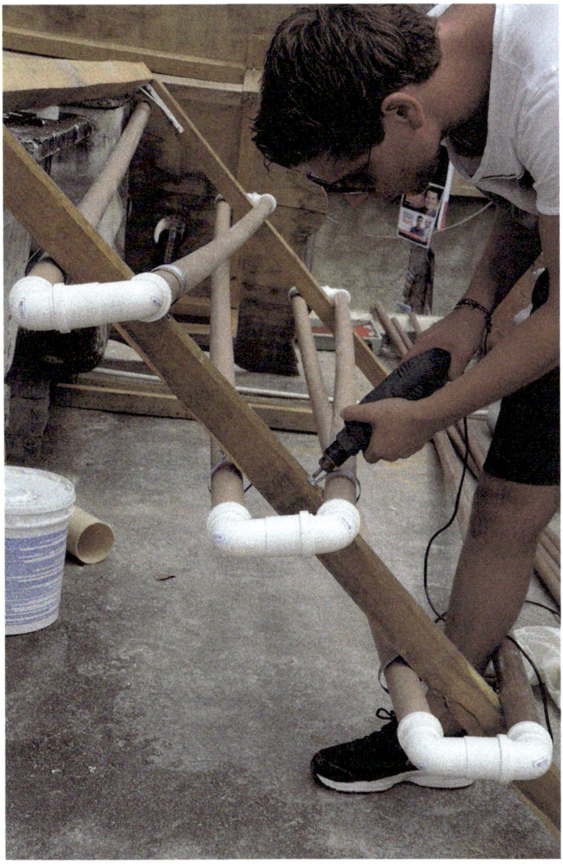

Fig. 9.25 Vertical system (*photo* Rob Roggema)

Before the plant-bed is nicely white and waterproof and the pipes are tied together it is dark in the favela, but the gleam from the surroundings makes the project look fairylike (Figs. 9.36 and 9.37).

By the end of a day of hard work sometimes you make stupid mistakes. How should the T-piece be glued again? Was it this way or the other way around? Once it is fixed it is too late. We need to buy a new one tomorrow and repair. More haste, less speed ....

The tilapia is on its way to Cantagalo and will be delivered tomorrow morning. Hopefully, by then the water supply is working again. Without water nor fish-tanks, no fish, no aquaponic system, and no FoodRoof. It is going to be tensive.

Fig. 9.26 Drilling holes for the cups (*photo* Rob Roggema)

9.6 Day Five

The first whats-app early morning: Marcelo Assunção confirms the favela receives water again and the tilapia has arrived. Except for the plants and the cups for the vertical system, we now have everything we need. The tanks can be filled up with water and later on the fish can swim.

When we arrive this morning we can see from far the ever more beautiful FoodRoof. The red-white-blue fish-tanks can be seen from distance and stand out in the environment (Fig. 9.38).

While Marc and Bart start to regulate the system, are we, Marcelo Mourão and driver Jorge, hunting for plants. The plants are, after waiting for the driver at metro-station General Osório, easy to find. It is a bit of a drive, but we are rewarded

Fig. 9.27 The vertical system stands (*photo* Rob Roggema)

with 10 plants (bell pepper, chilli pepper, tomato and strawberry) (Fig. 9.39) and seven pots with herbs, basil, rosemary and sage for in the vertical system. We purchase also 60 cups for in the vertical system, and buy the cactuses with it. Hop, everything in the back of the car (Fig. 9.40) and swiftly back to Cantagalo.

When we return to Cantagalo the feeder tanks are installed and the tubes connected (Fig. 9.42). Confident we attach the final tubes and pipes and the water can start to flow through the system. The tanks are full and the first drops of water enter the horizontal grow-bed. This wooden construction, earlier in the week carefully kitted and made waterproof, contains of a tiny hole here and there. While the circulation of the water runs very well, we discover small leakages from the grow-bed. The water runs too well…. The horizontal grow-bed leaks. A lot. Now, the entire roof is flooded instead of the grow-bed only. Only a couple of hours before a large group of people are invited to attend the opening of the FoodRoof, the cistern is leaking….

Fig. 9.28 Wood for the horizontal grow-bed arrives (*photo* Rob Roggema)

Fig. 9.29 Towing up the wood (*photo* Rob Roggema)

Fig. 9.30 manufacturing the horizontal grow-bed (*photo* Rob Roggema)

We decide to put a plastic layer at the bottom of the grow-bed. This could be quickly bought at the local shop, where the owner starts to smile more and more every time we show up....

The waterproof plastic is put in the cistern, we create a hole for the siphon and kit the edges completely (Fig. 9.41).

Finally the fish-tanks are definitely connected with pipes and tubes to the horizontal and vertical garden (Fig. 9.42).

The herbs can be put in the vertical system (Figs. 9.43 and 9.44).

The horizontal grow-bed can be filled up with clay balls (Fig. 9.45) and the water system can be started again (Fig. 9.46). The cistern is no longer leaking. The system works. Only the siphon needs to overflow, which occurs one hour later.

Fig. 9.31 The horizontal grow-bed gets shape (*photo* Rob Roggema)

The system is in operation. No leakages anymore and the crops can be planted in the horizontal grow-bed (Fig. 9.47).

Everything is ready and the roof is cleaned up. Time to release the fish in the fish-tank. The tilapia quickly swims its first laps (Fig. 9.48). The first FoodRoof is complete!

At two o'clock the opening starts. There is a lot of attendance: The Dutch consul, Arjen Uijterlinde and his wife, several Brazilian building companies, people from EMOP, the presidents of the community groups, residents, interested people from Mangueira and Babilonia, Amaury and the owner of the house, Marcelo Assunção, are all there (Fig. 9.49). After a brief speech and explanation of the system,

Fig. 9.32 Waterproofing the horizontal grow-bed (*photo* Rob Roggema)

simultaneously translated, the response is enthusiastic and supportive for the first FoodRoof. People are very positive and see opportunities to realise this type of system on other roofs in Cantagalo and in other favelas in Rio de Janeiro. In the meantime the FoodRoof wasn't unnoticed. Many residents of the favela were very curious. Marcelo's house is located alongside a busy street and the building activities received attention during the entire week, but once the FoodRoof is ready, one after the other came along to have a close look. Muito Bom! This is a roof I would like to have!, is an often received response.

To make some extraordinary pictures from above, we undertake an adventure and climb up the opposite roof. The ladder is half a human being too short to reach the upper roof, but once you've overcome this tiny problem, the only problem

Fig. 9.33 Fabricating the siphon (*photos* Rob Roggema)

remains is getting back down. Even this adventure ended successfully and the pictures turned out to be beautiful (Figs. 9.50, 9.51 and 9.52)!

The FoodRoof is complete. And is embedded beautifully in Cantagalo (Fig. 9.53).

The opening is closed with a Brazilian brunch (Fig. 9.54), made by the owner of the house, Marcelo Assunção. We gave him extensive explanation how to regulate and maintain the FoodRoof and what he needs to do in the upcoming weeks to keep the system in operation.

It is a unique experience to see the FoodRoof been build up from the very first t-piece until the system completely works. A whole process of support, advancements, barriers, problems and disappointments required adaptation and flexibility to the situation. It was a instructive and exiting process. The goodwill we have achieved in Cantagalo means a lot as well is the perspective of eating healthier food. To go short: a wonderful building week with a happy end.

Fig. 9.34 Connecting the pipes to the fish-tanks (*photo* Rob Roggema)

Fig. 9.35 Fitting the distribution pipes to the horizontal grow-bed (*photo* Rob Roggema)

Fig. 9.36 Waterproof horizontal grow-bed and vertical system (*photo* Rob Roggema)

Fig. 9.37 Vertical system on the roof (*photo* Rob Roggema)

Fig. 9.38 The FoodRoof in the morning (*photo* Rob Roggema)

Fig. 9.39 Looking for plant material (*photo* Rob Roggema)

Fig. 9.40 Plant material in the back of the car (*photo* Rob Roggema)

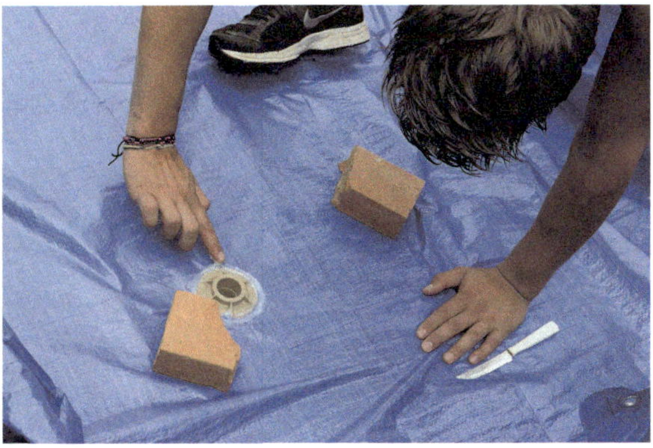

Fig. 9.41 Sealing the siphon (*photo* Rob Roggema)

Fig. 9.42 Connecting feeder tanks, tubes and water pump (*photo* Rob Roggema)

Fig. 9.43 Planting the plants in the cups (*photo* Rob Roggema)

Fig. 9.44 The vertical system (*photo* Rob Roggema)

Fig. 9.45 Filling the horizontal grow-bed with clay balls (*photo* Rob Roggema)

Fig. 9.46 Connecting the fish-tanks with the horizontal grow-bed (*photo* Rob Roggema)

Fig. 9.47 The horizontal grow-bed planted (*photo* Rob Roggema)

Fig. 9.48 Tilapia in the fish-tank (*photo* Rob Roggema)

Fig. 9.49 The FoodRoof is ready (*photo* Marcelo Maurão)

Fig. 9.50 FoodRoof from above (*photo* Rob Roggema)

Fig. 9.51 Foodroof from above (*photo* Rob Roggema)

Fig. 9.52 FoodRoof with a view over Cantagalo and Ipanema (*photo* Rob Roggema)

Fig. 9.53 The FoodRoof embedded in Cantagalo (*photo* Rob Roggema)

Fig. 9.54 Celebrating the opening of the FoodRoof (*photo* Rob Roggema)

Reference

Broekhuis, B., & Drissen, M. (2014). *Food growing roof terrace; Instruction manual aquaponic system, roof terrace Rafael Lezinho*. Velp: Van Hall Larenstein.

Chapter 10
Interview with Marcelo Assunçao, Owner of the First FoodRoof

Rob Roggema and Ruth Jurberg

10.1 Introduction

Marcelo Assunçao is the owner of the first FoodRoof. He is living in the heart of Cantagalo. His house is next to the busy main road of the favela where many residents pass. He is also a member of the community group, which is involved in commenting, participating and collaborating in the PAC program. He owns the house of five storeys and lives there with his wife and children, and his parents. He exploits a bed and breakfast, and finds it therefore very important how his house is presented. He is also keen to increase the sustainable exposure of his property and the FoodRoof is a welcome addition to realising this ambition. In this chapter we have interviewed Marcelo about how he became involved in the project, what his experiences are before, during and after the FoodRoof was built, and how the FoodRoof is operated.

10.2 Interview

10.2.1 Before the FoodRoof Was Built

How did you hear about this project for the first time?
Marcelo (Fig. 10.1): 'I am interested in the subject of urban green and food for at least 10 years. I tend to watch a TV show, which presented in one of its editions a

R. Roggema (✉)
Faculty of Design, Architecture and Building, University of Technology Sydney, Ultimo, NSW, Australia
e-mail: rob@cittaideale.eu

R. Jurberg
State Government of Rio de Janeiro, Rio de Janeiro, Brazil

R. Roggema (ed.), *Food Roofs of Rio de Janeiro*,
DOI 10.1007/978-3-319-56739-6_10

Fig. 10.1 Marcelo Assunção

system that would be the solution to urban life and gardening at home without the use of soil. Therefore I knew this system existed and when the people of the PAC program approached me with the question whether I would be interested I already understood its implications and opportunities it offered'.

Who told you about the FoodRoof?
Marcelo: 'I have a colleague who was giving technical support to a group of residents at Cantagalo community in a cultural fair. She was looking for a place that would accept a proposal for a food-growing roof. In a recent visit to Cantagalo she

came across my house, of which the garden was compromised by the public works been carried out. Therefore she had no doubt in asking me whether I wanted to participate in the FoodRoof project'.

What did they tell you?
Marcelo: 'I was introduced to the idea and the possibility of receiving the FoodRoof project. She subsequently asked me if I was open to a visit of an engineer/architect, who would carry out an analysis of my house and two other houses, whether the structure of the house was strong enough to carry the proposed FoodRoof system'.

What was your opinion when you first saw the proposal? Were you curious, did you think it might be a business opportunity?
Marcelo: 'In principle, I realised this was an opportunity to share an innovative project with my neighbours and friends. At first instance I had no intention to market the result or see it as an economic benefit'.

What do you expect in terms of benefits?
Marcelo: 'The main benefits I'd expected were to share the project results and motivate other interested participants to develop a similar system. Moreover, the realisation of the first FoodRoof could draw attention to the importance of green in the urban landscape. Finally, I thought the project could potentially promote a healthy lifestyle through this project'.

Did you get all the information you needed? What kind of information was not given to you?
Marcelo: 'I received the printed manual, explaining how to handle and I have followed the installation of the system. It was very significant to follow the installation step-by-step so that I could better understand the operation of the system and be able to solve potential future problems when they might occur'.

Why did you accept having the project in your home?
Marcelo: 'As stated earlier, the project is a great opportunity to multiply values of green landscape and healthy living, and especially to point out the importance of being able to produce food at home'.

10.2.2 During the Building of the FoodRoof

What do you think of the manual, was it helpful?
Marcelo: 'The manual (Fig. 10.2) was a clear and complete guide, which has been thoroughly elaborated to include all the necessary information for the reproduction of the system. There's no one who does not get carried away by the way everything is presented in the manual'.

Were there any problems you have encountered during the building process?
'However, some of the materials were difficult to find because of the short period that was available for completion of the project. Many components needed to be

Fig. 10.2 Manual for building a FoodRoof in Portugese (Broekhuis and Drissen 2014)

replaced by alternatives and of course, these adjustments required more time, which occasionally resulted in some problems'.

Did you encounter any other problems?
'The weather was another impediment, because during the days the system was realised the wind chill was 55 °C. All of these factors hindered the functionality and expected result of the system'.

How did the collaboration go with the people who were installing the FoodRoof?
'All the people involved in the construction were very attentive and did everything possible to construct the system in time and made sure the system was working. On the fifth day in the morning the installation was finished. Early afternoon several guests were coming to officially open the FoodRoof. These people were very satisfied and excited about the end result. Everyone was very optimistic about possible system improvements'.

10.2.3 After the Construction

How did the operation go in the period after installation? Did any problems occur?
'In the first two weeks after realisation we tried to maintain a constant cooling of the system in order for the fish to survive the terrible heat of the moment. However, we were not very successful and the fish died. The unexpected and severe summer heat also impacted the feeder tank and PVC pipes. During the first three days of implementation of the system, there were problems in the permeability of the feeder tank. The water was not easily flowing to and from the feeder tank, so not enough water reached the plant-bed'.

Was there anything else not functioning as expected?
'The total autonomy of the system was compromised by not having installed solar panels, responsible for generating electricity for the operation of the pumps'.

What were the main reasons for the system to not function as expected?
'It was very unfortunate the system didn't operate for a longer period, as we had expected. When the fish died we had to decide to deactivate the system, until the intensity of the rays of sun was reduced and the system wasn't suffering anymore from extreme heat. Because the feeder tank was built of wood, it could not withstand the strong heat, which caused leaks and made it impossible to operate'.

Would recommend this system to others?
'Despite the problems mentioned above, it is well worth to invest in this project, which can provide awareness of the importance of growing healthy food to the people. There are numerous benefits that the system provides, of which one of the most important is the direct improvements it could have on people's health'.

Which components of the system work well and which not?
'Pipelines, pumps, expanded clay-balls, filters and water distributors are easily available for this system. Other components, such as the feeder tank, which was made of wood and the hoses have been deteriorated over time'.

Did you gain a lot of attention for the FoodRoof?
'The project was very well publicised. It generates numerous visits from people who are interested in getting to know the system and wanting to see the system, amongst them many tourists pay us with a visit. Therefore it is imperative that the FoodRoof exists real time, so people have access to the system, can see the idea and also want to produce their own food'.

10.3 Conclusion

Marcelo is, in general very positive about the whole process. He is a bit disappointed by the short period the system actually operated. If he receives some help, especially with the materials that made the system malfunction, such as the feeder tank, which could have been better made from plastic and the hoses, he is keen to re-install the system and maintain it.

It is also necessary to have a pilot-system up and running, because of the numerous people that want to come along and have a look at the system. It seems the publicity in different media has sparked interest for the FoodRoof concept. However, in order to prove the functionality of the system to these visitors, the system must be in operation.

To withstand the heat the FoodRoof requires a cover, excluding excess heat from the plants and some of the components of the system. When this cover could be (partly) constructed with solar panels the required energy for the pumps could be generated within the system as well.

Marcelo is very supportive of the underlying purpose and principles of the FoodRoof. He would like others to obtain the benefits of having a FoodRoof also and would be happy to support anyone who wants to realise one.

Reference

Broekhuis, B., & Drissen, M. (2014). *Horticultura em terraços. Manual de instruções para sistemas de Aquacultura, Roof Terrace Rafael Lezinho*. Velp: Van Hall Larenstein.

Author Biography

Ruth Jurberg Urban Planner, working at the State Government of Rio de Janeiro since 1983. General Coordinator of PAC Program since 2008.

Part III

Chapter 11
Conclusions and Outlook

Rob Roggema and Ruth Jurberg

11.1 Conclusions

In this book we have presented the FoodRoof in Cantagalo, Rio de Janeiro. The first FoodRoof has been realised, but this is by far not enough to solve the access to food for poor people. The food problem is a global problem and the main answer to this problem lies at the city-region level. For the city of Rio de Janeiro this implies working at the distribution of food as well as the places where food is produced. The distances between supply and demand need to be shortened and especially for favela families it is important the supply comes from within the favela. Therefore this first initiative of the FoodRoof deserves followers of the example.

The FoodRoof will not solve the global food problem. However it will be able to solve the local food problem hence contributing to solutions at regional and global levels. Once the majority of the favela inhabitants could provide their own food by producing enough vegetables and fish for every single family, a huge step forward is made in reaching out to the people with the biggest distances to access healthy and enough food. Therefore the amount of food produced in the urban neighbourhoods, such as Cantagalo, needs to provide food for the entire family living under one FoodRoof. In addition, the entire favela could provide in its own demand.

The role of the FoodRoof extends beyond the provision of food. It also creates a community feeling within the favela. A FoodRoof makes it possible, even necessary to exchange crops during the year. When some family has a surplus of one crop it can be exchanged for other surpluses of other families. Social coherence is

R. Roggema (✉)
University of Technology Sydney, Sydney, Australia
e-mail: rob@cittaideale.eu

R. Jurberg
State Government of Rio de Janeiro, Rio de Janeiro, Brazil

R. Roggema (ed.), *Food Roofs of Rio de Janeiro*,
DOI 10.1007/978-3-319-56739-6_11

also improved through capturing the FoodRoofs as an element you would like to take care of. The produce is not expensive, and vulnerable at the same time, which makes it an object people would like to cherish.

The economic benefits of the FoodRoofs are evident. If you can have part of your meals from your own roof you do not have to buy all your basic needs in terms of food from the store. Also, whenever you produce more than you need you can exchange or sell the surpluses.

Finally the FoodRoof bring sustainability. The roofs require water and energy to becoming self-sufficient. This means rainwater must be captured and stored and solar energy should be the source of local energy consumption. Not only can this water and this energy be used for the FoodRoofs, but also for washing, showering, and to power electrical installations in home. Further the FoodRoofs are green year round, which gives a cooling effect on the entire area. Fears for the impacts of urban heat island effects (UHI's) are no longer necessary.

The realisation of more FoodRoofs in Brazil depends largely on the economic situation and the role the government would like to take. In economic difficult periods there is not much attention for greening the city. However, it would contribute economically in the sense that residents of favelas have more to spend when they produce their own food. Secondly, to build the FoodRoofs local workforce is required and it could be easily imagined to create local workshops of FoodRoof builders, in order to give young jobless people a change to learn and work.

The role the government could take up is to prescribe FoodRoofs as part of the bids for large building contracts, such as the PAC Program already planned to do. The social component is at the heart of the PAC Program. Within this social dimension, the education of environmental and sanitation principles is key, and the FoodRoof project is an excellent example of this.

When the FoodRoof is a regular part of the PAC program it becomes an inseparable component of rebuilding and refurbishing in the favelas, similar to giving every house three Fortlev water-tanks and connect them to the water supply.

11.2 Outlook

The first FoodRoof has been built. Looking into the future there are ample opportunities to extend the number of FoodRoofs. First of all the spread of FoodRoofs over Cantagalo should be enhanced as soon as possible. With the first good example available, it is relatively easy to create the next ones. In order to achieve this and start-up this process of additional FoodRoofs, a champion or leader of FoodRoofs need to be found. This is a person who knows how to build a FoodRoof, has enough daily contacts with the population of the area and can stimulate people to build their own FoodRoof.

When the amount of FoodRoofs increases in Cantagalo, it becomes a trade mark, and other favelas will follow. When the government pushes the realisation in other favelas the same way as they did in Cantagalo, the spread of FoodRoofs continues.

First the FoodRoof can be exported to other favelas in Rio de Janeiro, then to other cities in Brazil and finally to other cities outside Brazil. It may even become an export product of the Federal government of Brazil.

After realisation of the first FoodRoof it is important to build capacity for exchanging knowledge, facilitate new FoodRoofs and being available for questions and maintenance issues. An important role is here to play for architects and social workers to expand the success of this first project.

What if the FoodRoof becomes a regular feature for slum housing around the world. It could change the lives of people in slums everywhere. There is a business to take care of, people need to maintain their FoodRoof, but at the same time they eat well and produce a healthy diet for themselves and their children.

The look of slums will change as well. With a green roof on every house the entire slum area becomes green, with all the positive effects on the micro-climate, ecology and air quality. From Rio to Nairobi to Mumbai, cities become healthier and more beautiful.

Author Biography

Ruth Jurberg Urban Planner, working at the State Government of Rio de Janeiro since 1983. General Coordinator of PAC Program since 2008.